Second Edition

HANDBOOK OF PETROLEUM REFINING TECHNOLOGY
REFINING MADE EASY

This book is a comprehensive guide that simplifies the complex world of petroleum refining. It provides a clear and concise overview of the fundamental principles, processes, and technologies involved in refining petroleum. Written by industry expert, the book covers topics such as crude oil properties, process units, environmental considerations, and product specifications. It offers practical explanations, diagrams, and real-world examples to help you understand the refining industry.

Matthew E. Parkerson
3/3/2024

Handbook of
Petroleum Refining
Technology

Second Edition

Refining Made Easy

Matthew E Parkerson

U & T Best print

New York City, USA

PREFACE

Today, refiners are faced with the need to invest billions of dollars in equipment in order to comply with environmental regulations, which are frequently imposed by political mandates with little regard for the actual effects on the economy and the environment. Laws and regulations frequently change their guidelines. Refiners are reluctant to spend millions or billions of dollars on equipment that may no longer meet requirements when the units come on line because the design and construction of new processing units require several years of lead time. For the "short-term," a lot of effort is being put into developing reformulated fuels that have little effect on environmental degradation. We use the term "short-term" because laws have already been passed prohibiting the use of hydrocarbon fuels and allowing only fuels that are completely polluting in the next two decades. Solar, electric, and hydrogen are the only nonpolluting fuels mentioned at this time. The petroleum industry has only a short window of opportunity to recoup the substantial investment necessary to comply with the current legal requirements because of this. Companies that maximize the efficiency and experience of their engineers and scientists will undoubtedly be the ones to emerge victorious from this era.

For this version, I have taken the new ecological parts of the business into account, as well as the utilization of heavier unrefined oils and rough oils with higher sulfur and metal substance. This large number of models influence the handling choices and the handling gear expected in a cutting edge processing plant. The fundamental parts of current petrol refining innovation and financial aspects are introduced in an efficient way reasonable for prepared reference by specialized directors, rehearsing engineers, college employees, and graduate or senior understudies in substance designing. Reformulated fuels' place in the distribution of refinery products and their impact on the environment are also discussed.

In the **Second Edition** of the book, Handbook of Petroleum Refining Technology: Refining Made Easy, - the content has been thoroughly reviewed and improved. The new edition includes additional topics to make the course meaningful and complete. These topics include: Crude oil distillation and definition of refinery capacity, Vacuum distillation, Super fractionation, Absorption, Adsorption, Crystallization, Conversion, Naphtha reforming, Catalytic cracking, Polymerization and alkylation, Hydrocracking, Isomerization,

Visbreaking, Thermal cracking and coking, Purification, Sweetening, Mercaptan extraction, Clay treatment, Hydrogen treatment, Molecular sieves, Petroleum, products and their uses, Petrochemicals, Refinery plant and facilities, Petroleum refining processes, etc.

The book presents the fundamental aspects of current petroleum refining technology and economics in an organized manner suitable for reference by technical managers, practicing engineers, university faculty, and graduate or senior students in chemical engineering. It also discusses the role of reformulated fuels in the distribution of refinery products and their impact on the environment.

The material is organized using the case study method of learning. The book covers refinery operation, design, and evaluation and provides practical approaches to various refinery issues. I have extensive experience in refinery operation, design, and evaluation, and have incorporated the knowledge and insights of many contributors into this edition of the book.

ACKNOWLEDGEMENTS

I would like to take this opportunity to express my deep appreciation and gratitude to all those who have contributed to the creation and success of this book, Petroleum Refining Technology: Refining Made Easy, Second Edition.

First and foremost, I extend my heartfelt thanks to Matthew E. Parkerson, the author of this book. Matthew's expertise, knowledge, and dedication to the field of petroleum refining have undoubtedly been instrumental in providing you with a comprehensive and accessible resource. His meticulous research, insightful explanations, and commitment to simplifying complex concepts have made this book an invaluable asset to both students and professionals in the field.

I would also like to acknowledge the invaluable contributions of the technical reviewers who provided their expert insights and suggestions. Their meticulous attention to detail and invaluable feedback have significantly enhanced the quality and accuracy of the content presented in this book.

Furthermore, I extend my appreciation to the editorial and production team who worked tirelessly behind the scenes to transform the manuscript into a polished and well-structured book. Their dedication, professionalism, and commitment to excellence have played a vital role in ensuring the book's readability and overall quality.

I would also like to express my gratitude to the publishers and everyone involved in the marketing and distribution of this book. Their efforts have been instrumental in making this resource widely available to readers around the world.

TABLE OF CONTENTS

INTRODUCTION

The United States is an oil-rich country, and crude oil is the mainstay of the nation's economy. With the economy of the country being dependent upon energy production, though not fully dependent as the case may be, control over the national energy future is necessary. Petroleum use is a necessary part of the modern world, hence the need for stringent controls over the amounts and types of emissions from the use of petroleum and its products. Therefore, it is predictable that petroleum will be a primary source of energy for the next several decades. The challenge is the development of technological concepts that will provide the maximum recovery of energy from petroleum cheaply, efficiently, and with minimal detriment to the environment.

Pollution has been obvious for a long time, although the effects were not realized in the past. Gas flaring at refineries was a norm. There was insufficient awareness of the effects of waste products on human life and since there was no form of environmental protection, the system of waste disposal thrived. The capacity of the environment to absorb the effluents and other impacts of process technologies is not unlimited. The environment should be considered as an extremely limited resource, and the discharge of chemicals into it should be subject to severe constraints.

Indeed, the declining quality of raw materials, especially crude oil, and fossil fuels that give rise to many of the gaseous emissions of interest, dictates that more material must be processed to provide the needed fuels. In addition, the growing magnitude of the effluents from fossil fuel processes has moved above the line where the environment can absorb such process effluents without disruption. To combat any threat to the environment, it is necessary to understand the nature and magnitude of the problems involved. It is in such situations that environmental technology has a major role to play.

CHAPTER 1

1. LUBRICATING OIL

Lubrication is a process of inserting lubricant in between rubbing surfaces to control friction and/or to reduce wear on the surfaces. Lubrication is a major component of tribology. Tribology is the science and technology concerned with interacting surfaces in relative motion; it includes friction, lubrication, wear, and erosion. Lubricants may be liquids, solids, gases, or greases that are designed to minimize contact between rubbing surfaces to allow easy shear so that the frictional force opposing the rubbing motion is low.

Lubricating oils are formulated for virtually every type of machine and manufacturing process. Lubricating oils contain refined or synthesized base oils from animal, vegetable, or petroleum, and a variety of additives that improve their lubricating and other characteristics. The base stocks and the type and concentration of additives used for these oils are selected based on the requirements of the machinery or process being lubricated, the quality required for the use of the machinery, and government regulations.

1.1. Uses of Lubricating Oils

Table 1 shows a partial list of the key uses/applications of lubricating oils. Each of these oils has a unique set of performance requirements. These requirements include proper lubrication of the machine or process, maintenance of the quality of the lubricant itself, the effect of the lubricant's use and disposal, energy use, the quality of the environment, and on the health of the user. Lubricating oils have the following uses/functions:

- Reduce friction and wear by interposition of a thin liquid between moving surfaces
- Remove heat – conserve energy
- Keep equipment clean – increase equipment life
- Prevent corrosion
- Operator or user safety

Automotive lubricating oils	Industrial lubricating oils
gasoline engine oils	industrial gear oils
for passenger cars and light trucks	pneumatic tool lubricating oil
for heavy duty automotive and industrial service	high temperature oils
for piston engines in general aviation service	air and gas compressor oils
small 2-stroke and 4-stroke gasoline engines	for reciprocating compressors
for outboard motors	for rotary vane compressors
for scooters, mopeds, and motorcycles	for rotary screw compressors
for lawnmowers and small tractors	for refrigeration compressors
for chain saws and similar portable equipment	machine tool way oils
diesel engine oils	textile oils
for heavy duty trucks, agricultural and construction vehicles	steam turbine oils
for industrial cross-head and trunk piston diesel engines	hydraulic fluids
for railroad diesel engines	paper machine oils
for marine cross-head and trunk piston diesel engines	food machinery oils
gas engine oils	steam cylinder oils
gas turbine oils	**Metalworking fluids**
for aircraft jet engines in commercial aviation service	for metal cutting
for industrial gas turbine engines	for metal rolling
automatic transmission fluids	for metal drawing, forging, stamping, etc
gear oils	
for automotive manual transmissions	
for automotive differentials	

Table 1: Uses/application of lubricating oils

1.2   Properties of Lubricating Oil Base Stock

Base stocks are manufactured to specifications that place limitations on their physical and chemical properties, and these in turn establish parameters for refinery operations. Base stocks from different refineries will generally not be identical, although they may have some properties (e.g., viscosity at a particular temperature) in common.

The most important properties of lubricating oil are viscosity, viscosity index, pour point, oxidation resistance, flash point, boiling temperature, and acidity (neutralization number).

1.2.1 Viscosity

The viscosity of a fluid is a measure of its internal resistance to flow. Lubricating oil may be thin and free flowing or thick with a high resistance to flow depending on the service for which it is used. From a given crude oil, the distillation boiling range of the cut can be used to select the viscosity of a blending stock. The higher the boiling point ranges of a fraction the greater the viscosity of that fraction: the higher the viscosity, the thicker the oil and the thicker the film of the oil that clings to a surface.

Within a naphthenic or paraffinic type, base stocks are distinguished by their viscosities and are produced to certain viscosity specifications.

1.2.2 Viscosity Index

The rate of change of viscosity with temperature is expressed using viscosity index (VI) of the oil. For a given change in temperature, the higher the viscosity index, the smaller its change in viscosity. The viscosity index of natural oils ranges from negative values for oils from naphthenic crudes to about 100 for paraffinic crudes. Specially processed oils and chemical additives can have a Vis of 130 and higher. Additives, such as polyisobutylenes and polymethacrylic acid esters, are frequently mixed with lube blending stocks to improve the viscosity-temperature properties of the finished oils.

1.2.3 Pour Point

The pour point is the lowest temperature at which oil will flow under standardized test conditions. It is reported to be at 50F or 30C increments of the oil. For motor oils, a low pour point is very important to obtain ease of starting and proper start-up lubrication on cold days.

There are two types of pour points:

Viscosity pour point: This is reached gradually as the temperature is lowered and the viscosity of the oil increases until it will not flow under the standardized test conditions.

Wax pour point: This occurs abruptly as the paraffin wax crystals precipitate from the solution and the oil solidifies.

A related test is the cloud point. The cloud point is the temperature at which wax or other solid materials begin to separate from the solution.

1.2.4   Oxidation Resistance

The high temperatures encountered in internal combustion engine operation promote the rapid oxidation of motor oils. This is particularly true for oil that is exposed to piston heads where temperatures can range from 2600C to 4000C. Oxidation causes the formation of coke and varnish-like asphaltic materials from paraffin-base oils and sludge from naphthenic-base oils. Antioxidant additives, such as phenolic compounds and zinc dithiophosphates, are added to the oil blends to suppress oxidation and its effects.

1.2.5   Flash Point

The flash point is the temperature of the oil at which it momentarily flashes in the presence of air and an igniting source. The flash point of oil has little significance concerning engine performance and serves mainly to indicate hydrocarbon emissions or the source of the oils in the blend. Low flash points indicate greater hydrocarbon emissions during use.

1.2.6   Boiling Temperature

The boiling ranges and viscosities of oil fractions are the major factors in selecting the cut points for the lube oil blending stocks on the vacuum distillation unit. The higher the boiling temperature range of a fraction, the higher the molecular weights of the components and, for a given crude oil, the greater the viscosity.

1.2.7   Acidity

The corrosion of bearing metals is largely due to acid attack on the oxides of the bearing metals. These organic acids are formed by the oxidation of lube oil hydrocarbons under engine operating conditions and by acids produced as by-products of the combustion process, which are introduced into the crankcase by piston blow-by. Motor oils contain buffering materials that neutralize these corrosive acids. Usually, the dispersant and detergent additives are formulated to include alkaline materials, which serve as neutralizers to the acid contaminants. The neutralization number is used as the measure of the organic acidity of oil, the higher the number, the greater the acidity.

1.3   Lubricating Oil Base Stocks

The most important component of lubricating oil is its base stock. Petroleum fractions having an average volatility than that of gas oil form the base of various fractions of lubricating oils. Petroleum base stocks are hydrocarbon-based liquids, which are the major components (80% to 98% by volume) of finished lubricants, the remaining 2% to 20% being additives to improve performance. Base stocks usually have boiling ranges between 600°F and 1100°F at atmospheric pressure (some are lighter) and lube feedstock therefore comes from the high boiling region—the vacuum gas oil fraction and residue—of crude oil. Base stock boiling ranges may extend over several hundred degrees Fahrenheit.

Base oils are mixtures of paraffin (straight or branched chain hydrocarbons). Although often supplemented by additives, the base oil determines the flow characteristics of the lubricant, its oxidation stability (sludge and deposit-forming tendency), its volatility, and its corrosion potential. Most lubricants, including naphthenes (cycloparaffins) and aromatics (alkyl benzenes and multi-ring aromatics), typically contain 20 – 40 carbon atoms per molecule.

When refined by conventional separation processes, the type of base stock is crude-specific, meaning that paraffinic base stocks come from paraffinic crude oils, and naphthenic oils from naphthenic crudes. When refined by modern conversion processes, base stocks are less crude-specific, since these processes are capable of converting naphthenic and aromatic compounds to paraffin.

If paraffins predominate, the base stock is paraffinic. If naphthenes predominate, it is a naphthenic base stock. Paraffins and naphthenes are saturated, meaning that all of the carbon atoms in the hydrocarbon molecule are singly bonded to another carbon atom or a hydrogen atom. These lubricant base stocks are manufactured by the distillation of selected crude oils, followed by further refining of the lube oil distillates by conventional separation or modern conversion processes.

Naphthenic crudes were the first crude oil used for the manufacturing of lubricating oils because; they give appropriate freezing points without undergoing further treatment. As the demand for lubricating oils of higher viscosity index increases, paraffinic crude oil becomes the main source of base oil production. However, too paraffinic crude oil causes difficulty in dewaxing, as the cake thickness makes filtrations difficult, and lowers the oil output prohibiting washing.

1.3.1 Base Stock Classification

The American Petroleum Institute (API) defines a base stock as ''. . . a lubricant component that is produced by a single manufacturer to the same specifications (independent of feed source or manufacturer location); that meets the same manufacturer's specification; and that is identified by a unique formula, product identification number, or both..."(Lynch, 2008). ''A base stock slate is a product line of base stocks that have different viscosities but are in the same base stock grouping and from the same manufacturer''

''A base oil is the base stock or blend of base stocks used in an API-licensed oil''.

API has also established five base stock categories, classified according to saturate content, sulfur content, and viscosity index. The classification is shown in the table below.

Table 2: API Base stock categories (Willey, 2007)

API group	Saturates content, mass % by ASTM D 2007		Sulfur content, mass %[a]	VI by ASTM D 2270
I	< 90	and/or	> 0.03	$80 \leq VI < 120$
II	≥ 90	and	≤ 0.03	$80 \leq VI < 120$
III	≥ 90	and	≤ 0.03	$VI \geq 120$
IV	polyalphaolefins (PAO)			
V	All other base stocks not included in group I, II, III, or IV[b]			

Base oils with higher saturated content are generally more resistant to, and easier to protect against oxidation. They also have a higher viscosity index. Paraffinic oils, at the same saturation level, have a higher viscosity index than naphthenic oils. Sulfur compounds produce corrosive material when oxidized. Base stocks with a wide molecular weight range tend to be more volatile compared to those with a narrow range. Base Stocks in API groups 'I–IV' are paraffinic hydrocarbons. Naphthenic base stocks are in group V. Typically, the saturated content of group

II and III base stocks is >99%, and the sulfur content is <15 ppm (<0.0015%). Group IV stocks are 100% saturated and contain no sulfur. The viscosity indexes of group II stocks are typically 100–115, and those of groups III and IV are 120–140. Group IV base stocks have the best low-temperature flow characteristics because they contain no wax. Naphthenic stocks, with minimal wax contents, also have good low-temperature flow properties. They also have the lowest viscosity indexes. The oxidation stability of the paraffinic stocks improves with each group number, as do volatility and deposit, sludge, and soot control.

1.3.2 Synthetic Base Stock

The term, synthetic, is used to differentiate between base stocks made by conventional crude oil refining processes from those synthesized from other chemicals. Polyalphaolefins in group IV are still called synthetic base stocks, as are the diesters, polyol esters, polyglycols, etc, in group V.

1.3.3 Base Stock Manufacturing Processes.

Figure 1 lists some of the refining and conversion processes used to manufacture groups I, II, and III base stocks. Base Stocks in API groups 'I–IV' are manufactured in several viscosity grades.

Crude oil is first fractionated in an atmospheric distillation tower to produce light gases and fuel products. The residue, or bottoms, from the atmospheric tower, are then fractionated in a vacuum distillation tower to produce gas oil and lube oil fractions. The vacuum residue is separated with propane to produce asphalt and deasphalted cylinder oil.

1.4 Manufacturing of Lubricating Oil

The use of animal fats to reduce friction and wear and tear of mechanical parts has been the practice from time immemorial. However, since the availability of petroleum sources, lubricants are now manufactured using petroleum stocks. Today's lubricating oil is mainly composed of base hydrocarbon oil, lubricating base oil stock (LOBS), obtained from vacuum distillates after treatment in the refinery, with some additives to meet the requirements for its end use. Base Stocks may be manufactured using a variety of different processes, which include distillation, solvent refining, hydrogen processing, oligomerization, esterification, and refining. Refined stock shall be substantially free from materials introduced through manufacturing, contamination, or previous use''.

18

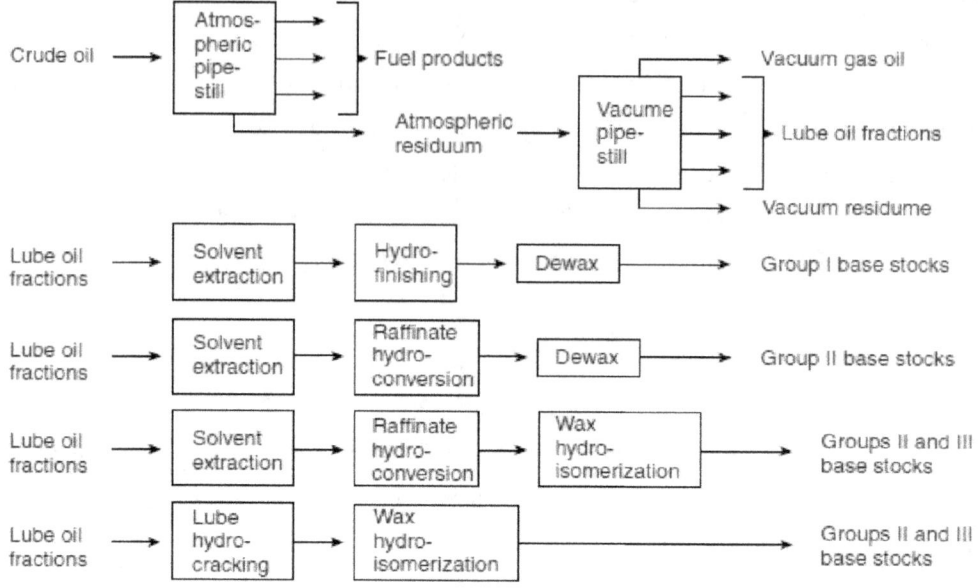

Figure 1: Base stock refining options (Wiley, 2007)

The first step in the processing of lubricating oils is the separation on the crude oil in distillation units to the individual fractions based on their viscosity and boiling ranges. The heavier lube oil raw stocks are included in the vacuum fractionating tower bottoms with the asphaltenes, resins, and other undesirable materials.

The raw lube oil fractions from most crude oils contain components that have undesirable characteristics for finished lubricating oils. These components must be removed to improve the oil quality. The undesirable characteristics include high pour points, large viscosity changes with temperature (low VI), poor oxygen stability, poor colour, high cloud points, high organic acidity, and high carbon- and sludge-forming tendencies.

The processes used to change these characteristics are (Lynch, 2008; Prasad, 2010):

1. *Vacuum distillation* – isolates individual raw lube oil fractions based on the desired starting viscosity

2. *Solvent deasphalting* – to reduce carbon- and sludge-forming tendencies, remove asphalt from vacuum residue, and produce a feedstock for very heavy lube oil called bright stock.

3. *Solvent extraction* - to improve viscosity index

4. *Solvent dewaxing* – to lower cloud and improve pour points

5. *Clay treating / Hydrofinishing* - to improve colour, colour stability and oxidation stability

Although the main effects of the processes are as described above, there are also secondary effects. For example, although the main result of solvent dewaxing is the lowering of the cloud and pour points of the oil, solvent dewaxing also slightly reduces the VI of the oil. For economic and process reasons, the process sequence is usually in the order of deasphalting, solvent extraction, dewaxing, and finishing. However, dewaxing and finishing processes are frequently reversed. In general, the processes increase in cost and complexity in this same order. Figure 2 is a diagram of a typical lube oil manufacturing complex in the refinery.

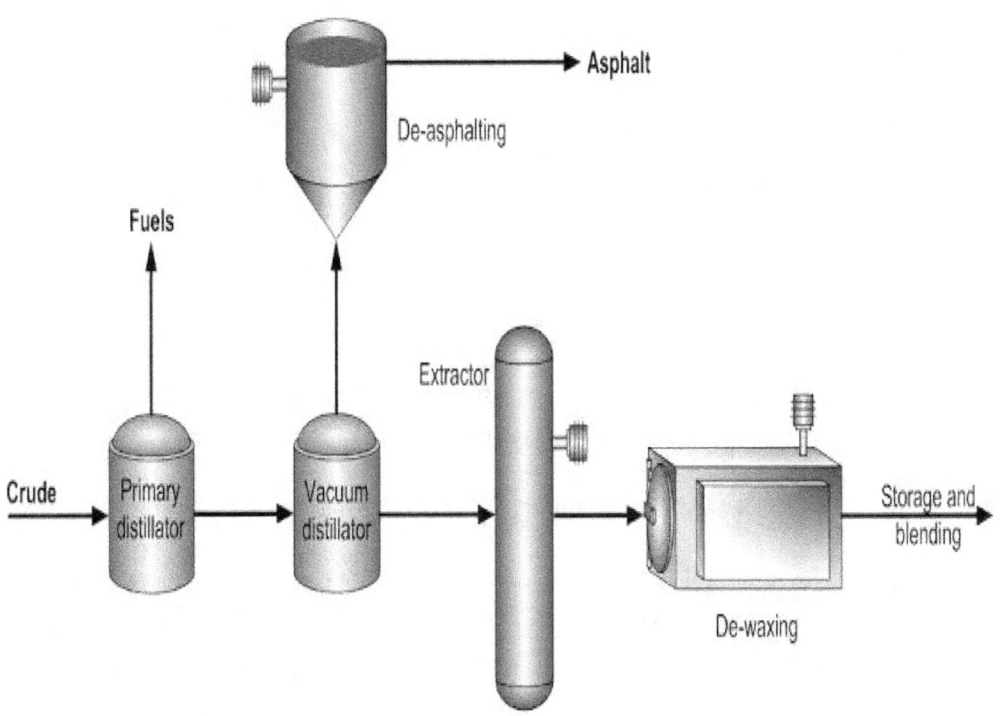

Figure 2: Diagram of lube oil manufacturing processes.

1.4.1. Vacuum Distillation

In vacuum distillation of long residue, the feed is spilled into heavy waxy distillate fractions such as spindle oil, light oil, intermediate oil, heavy oil cuts, and short residue. The long residue fractionation is achieved by distillation under an appropriately high vacuum to prevent the decomposition of very high boiling components. Figure 3 depicts the process.

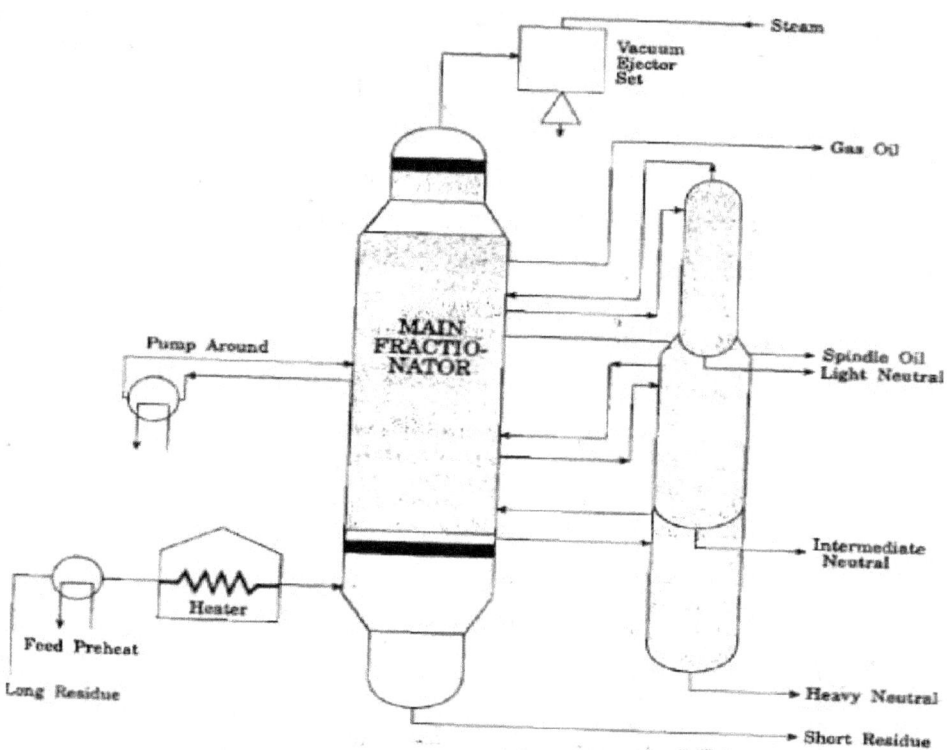

Figure 3: Vacuum distillation for lube distillates production

1.4.1.1   Process Description:

The long residue from the atmospheric column is heated further, vaporized in the fired heater, and then charged to a vacuum column. The operating temperature and absolute pressure are 390-400°C and 120mmHg respectively. The required vacuum is maintained by a multistage steam ejector system. The partial pressure of hydrocarbons in the vacuum column is reduced effectively by the injection of steam at the bottom of the column. The distillate fractions are steam stripped in the strippers to control the flash points. These fractions are then sent for further processing.

Typical operating conditions:

- Flash zone temperature: 385^0C
- Absolute pressure: (a) Top – 23 mmHg, (b) Flash zone 100 mmHg

1.4.2 Solvent Deasphalting

Solvent deasphalting process is used to remove asphalt from vacuum residues, in other to produce very heavy lubricating oils called bright stocks. The most commonly used solvent for deasphalting, which gives better yield and high-grade refined product is propane. However, other light hydrocarbons are now used either alone or in combination with propane, depending on the nature of the feedstock and the quality of the deasphalted oil – to optimize yield and operating cost. Propane extracts the high quality oils in the vacuum residues by precipitating the asphaltenic and resinous components containing metals, sulphur and nitrogen.

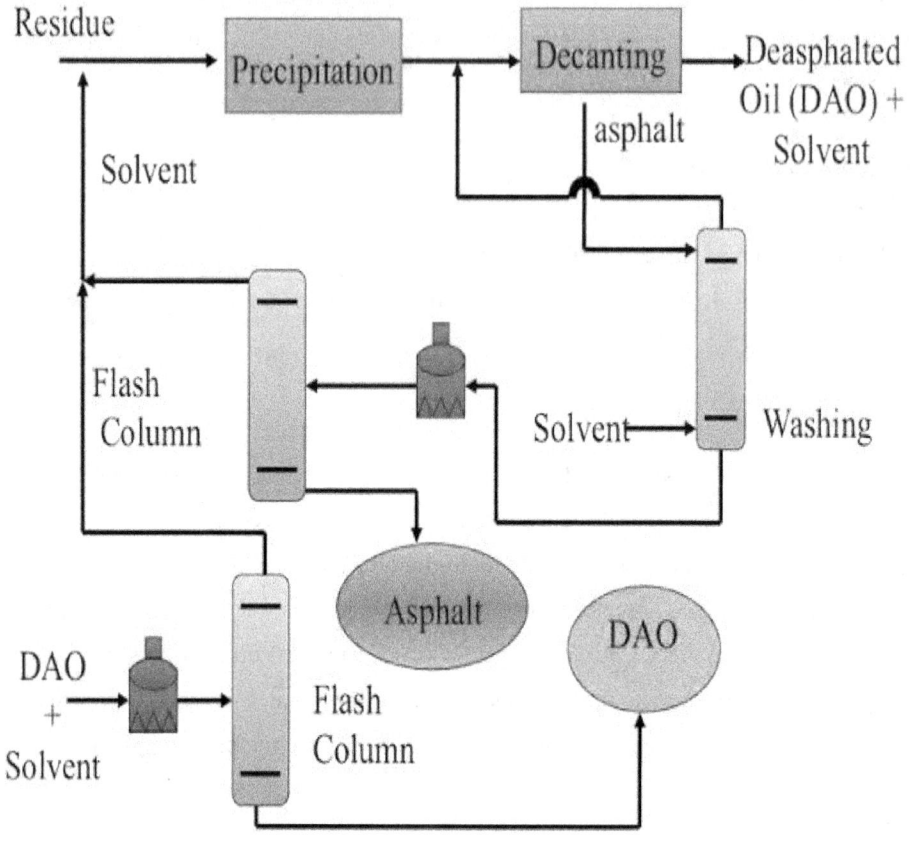

Figure 4: Schematics of a deasphalting process

1.4.2.1 Process Description:

Figure 4 is a simplified flow diagram of a deasphalting process. Vacuum residue (the feed) is heated to the specified temperature and then charged to the top of the extractor for propane.

1.4.3. Solvent Extraction

In the solvent extraction process, lube oil feedstock from the vacuum tower flows upward through a treating tower, counter-current to a stream of solvent. The solvent preferentially removes undesirable tars, resins, asphaltic compounds, and polycyclic naphthenes, and aromatics. Solvents used include phenol, furfural, and N-methyl-pyrrolidone (NMP). The solvent is then stripped from both the aromatic extract and raffinate streams. Solvent extraction increases the VI and the stability of the raffinate.

1.4.4. Dewaxing

Paraffinic distillates contain some high molecular weight and high melting point paraffin waxes. At temperatures below their melting point, these waxes crystallize and cause the oil to gel. Several methods are used to reduce the wax content of lubricating oil base stocks. In one method, the hydroformed raffinate is mixed at a low temperature with a chilled solvent, in which the heaviest waxes are insoluble. The wax crystallizes and is then separated from the mixture by cold filtration. Solvents used include propane, methyl ethyl ketone (MEK), methyl isobutyl ketone, and a mixture of MEK and toluene.

Another method for removing wax from the raffinate is catalytic dewaxing. This method uses a shape-selective dewaxing catalyst to crack straight chains or slightly branched waxes into naphtha and gas.

Isomerization technology is used in the manufacture of group II and III base stocks. In this process, high melting point straight-chain paraffins are isomerized in a shape-selective catalyst to branched-chain, lower pour-point lube molecules. The process also produces gasoline and diesel fuels.

1.4.5. Finishing

The essence of the finishing process for the refining of lube base stocks is to improve color stability and performance by removing some of the remaining heteroatoms (, higher molecular weight sulfur-, nitrogen- and oxygen-containing compounds) which are among the more easily oxidized components during lubricant use and contribute to the formation of sludge, color, and oxidation products. Although most of these compounds have been removed in the extraction stage, some may remain and are handled in the finishing process. There are two processes, that have been employed, they are:

I.   Clay-treating, which separates them by adsorption, and

II.   Hydrofinishing converts them into acceptable lube components by hydrogenation.

1.4.5.1   Clay Treating

Clay treating is a high-temperature and pressure process used to heat naphtha to improve the stability and color of lubricating oil. The stability is increased by the adsorption and polymerization of reactive diolefins in the cracked naphtha. Clay treating is also used for treating jet fuel to remove surface-active agents, which adversely affect the water separator index specifications.

The clay treating process penetrates base stock through a heat-activated solid adsorbent, usually bauxite (a form of aluminum oxide; e.g. Porocel) or naturally occurring clay (Fuller's earth, Attapulgis clay). Polar impurities such as nitrogen, oxygen, and sulphur-containing compounds are adsorbed on the solid surface and removed from the oil. Figure 5 provides a general schematic for the process.

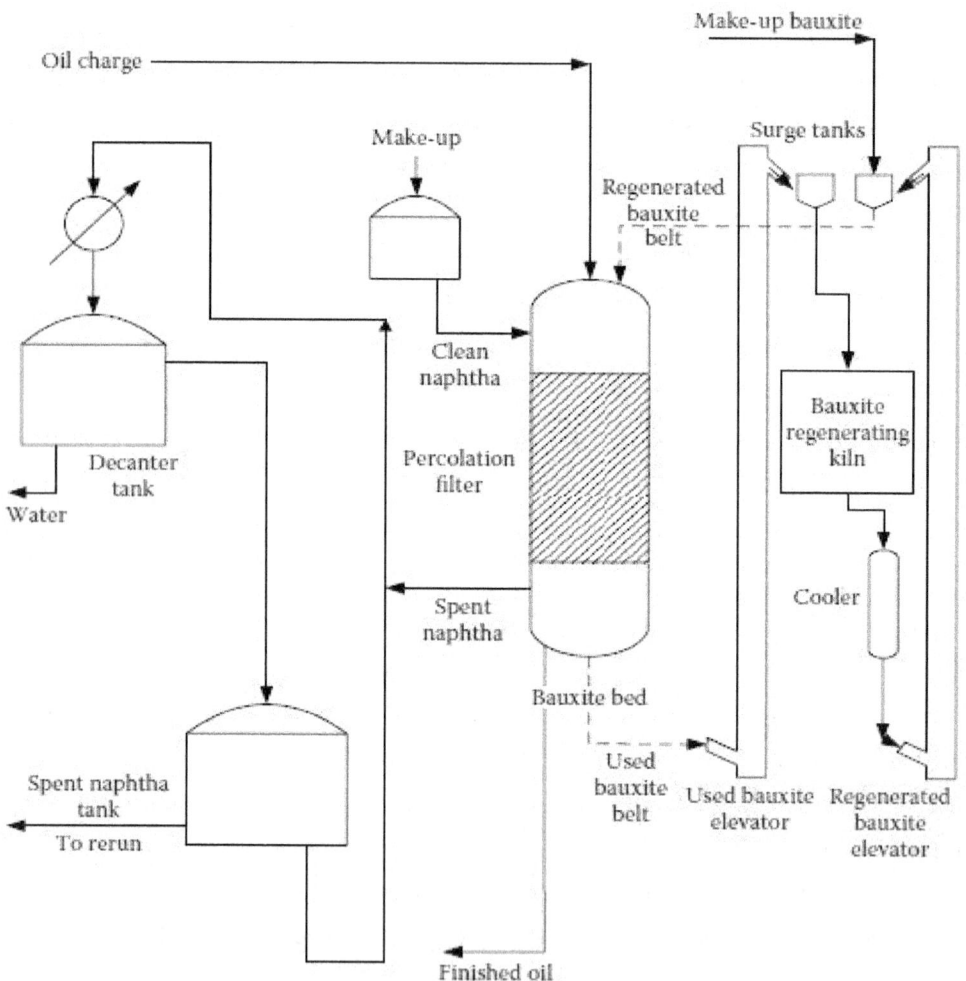

Figure 5: Process flow diagram of a clay treating unit

Heated oil charge is fed to a bauxite bed and allowed to percolate through at a predetermined rate. When specifications can no longer be met, the clay is declared exhausted and charge feed is discontinued. A flush with naphtha removes the remaining oil and some of the adsorbed components, then a steam purge is employed to remove the naphtha and the clay can be transferred to a belt that carries the hydrocarbon-free clay to the regenerator kiln to burn off the remaining adsorbed material. This step also adjusts the water content of the clay to the correct level, which is needed since water level affects adsorption. The regenerated material is then returned to the percolation filter bed and reused. This process has also been employed as part of the traditional route for purifying waxes and as a clean-up method for hydrotreated white oils.

1.4.5.2 Hydrofinishing

Hydrofinishing is catalytic hydrogenation process that converts unstable compounds remaining after solvent extraction to stable ones. It converts some aromatics to naphthenes and removes some sulphur compounds and other trace materials. Compared to the modern hydrocracking process that causes major molecular changes, hydrofinishing is a relatively mild process. However, it is severe enough, to produce naphthenic base oils from unextracted distillates that do not require labelling as carcinogenic under OSHA's Hazardous Substances Communication Standard.

1.5 Hydrogen Processing

Hydrogen reforming, hydrocracking, and wax isomerization are modern refining processes that convert undesirable components of lube oil fractions into desirable components, rather than remove them. They are used instead of solvent extraction on vacuum tower distillates, or on raffinates from the extraction tower, to produce group II and III lubricating oil base stocks with higher saturates content, lower sulphur content and higher viscosity index. The principal reactions taking place in these processes are saturation, ring opening, reforming (isomerization), cracking, desulfurization, denitrogenation.

Naphthenic and group I paraffinic base stocks are refined by conventional separation processes and are crude specific. Groups II and III paraffinic stocks produced by modern conversion processes, hydrogen reforming, hydrocracking, catalytic dewaxing, and wax isomerization, are less crude specific.

1.6   Polyalphaolefins

Group IV base stocks are poly (a-olefins) (PAO). They are produced by steam cracking hydrocarbons to produce ethylene; ethylene oligomerization to produce linear a-olefins (1decene, 1-dodecene, or 1-tetradecene); oligomerization of linear a-olefin to form a mixture of dimers, trimers, tetramers, and higher oligomers; hydrogenation of the unsaturated oligomer.

Hydrocarbon Steam cracking $\longrightarrow$ Ethylene Oligomerization Linear a-olefins Oligomerization   Dimers Trimers Tetramers

dimers   trimers, tetramers and higher oligomersPAO   Higher oligomers

The chain length of the olefin raw material, temperature, time and pressure, catalyst types, and concentration and distillation of the final product to remove the dimers affect the characteristics of the finished PAO. The conversion of natural gas to liquids (GTL) by the Fischer Tropsch process is a promising method for producing high-quality lubricating oil base stocks.

Organic Esters: Organic esters synthesized by reacting dibasic acids with mono alcohols (diesters), or by reacting monoacids with polyhydric alcohols (polyol esters), have been used as lubricating oil base stocks for>50 years. They are branched hydrocarbon molecules with multiple ester linkages, which give them polarity. Their unique structure gives them excellent thermal and oxidation stability, good low-temperature flow characteristics, low volatility, lubricity, detergency and dispersant, and biodegradability. Esters have been used exclusively in aircraft turbine engine oils (jet engine oils) for>40 years. They are also the preferred base stock in refrigerator compressor lube oils used with R-134 refrigerants. Other applications for ester base stocks include rotary screw air and process gas compressors, oven chain lubricants, and gas engines.

Polyglycols: Polyalkylene glycols (PAG) and polyethers are usually copolymers of ethylene oxide and propylene oxide. The oxide monomer sequence can be random or blocked and their solubility can be varied from oil-soluble to completely water-soluble. Their applications include gear and compressor lubricants, metalworking fluids, fiber lubricants, and fire-resistant hydraulic fluids.

Vegetable Oil Esters: Triglyceride esters are obtained from renewable sources: olive, soybean, rapeseed, canola, safflower, sunflower, meadowfoam, castor, and other vegetable oils. They are biodegradable and offer specific environmental benefits over hydrocarbon-based lubricants. Lubricating oils made with these base stocks and their derivatives are recommended in applications where lubricant leaks and spent lubricant can contaminate the environment. Examples include chain bar lubricants, two-cycle oils for outboard marine engines, hydraulic fluids for farm machinery, and rail curve greases.

Most vegetable oil esters have a combination of saturated and unsaturated fatty acids attached to the three alcohol groups in glycerine. Highly saturated oils have good oxidation stability and poor low-temperature flow properties. As the amount of saturation decreases, oxidation stability decreases and the low-temperature flow properties improve. Advances in breeding technology can change fatty acid profiles and alter the physical properties of vegetable oils.

Biodegradable Base Stocks: Lubricant or base stock biodegradability is the extent to which the material can be broken down by living things (microorganisms) into innocuous products such as water, carbon dioxide, inorganic compounds, and methane. The least biodegradable lubricant base stocks are silicone oils, polyphenyl ethers, perfluoro alkyl ethers, and alkylated aromatic oils. Naphthenic stocks and base stocks in API groups I–IV also have relatively poor biodegradability. The most biodegradable base stocks are vegetable oil esters, followed by polyethylene glycols, organic esters, and phosphate esters.

Other Base Stocks: Other chemical compounds used as lubricant base stocks include polybutenes, hydrocarbons obtained by the alkylation of naphthalene with a-olefins, alkylated aromatic hydrocarbons, silicones, phosphate esters, chlorotrifluoroethylene, polyphenyl ethers, perfluoroalkyl polyether.

Rerefined Base Stock: In its definition of a base stock, the API states ''Rerefined stock shall be substantially free from materials introduced through manufacturing, contamination, or previous use''. Products made from re-refined stock are subject to the same stringent refining, compounding, and performance standards applied to virgin oil products. Refined oil may have superior oxidation stability than virgin stocks because the easily oxidized compounds will have been reacted during previous use and then removed during reprocessing.

In the refining process, used oil is preferably segregated by type, collected, and delivered to the reprocessing facility. The oils are then screened and inappropriate feedstocks are rejected. Solid and other gross contaminants are then separated by, eg, propane precipitation, alcohol– ketone precipitation, acid–clay filtration, etc. The filtrate is dehydrated and then fractionated by vacuum distillation. Hydrotreating or clay filtration or both then finish the distillates, and then vacuum distilled to obtain the desired viscosity grades.

1.7   Additives for Lubricating Oil and Grease

Advanced refining and manufacturing technologies have significantly improved the quality of base oils used in the manufacturing of lubricating oils and greases. They are more stable, have better low-temperature flow properties, are less volatile, and less corrosive. However, they do not meet the requirements of modern lubricating oils and greases. Additives are chemical substances added to the base oil to impart or improve certain properties. They include oxidation inhibitors, rust and corrosion inhibitors, pour point depressants, viscosity (VI) improvers, detergents, dispersants, friction modifiers, EP agents, thickeners, emulsifiers, demulsifiers, bactericides, fungicides, and tackiness additives.

Most oils and greases contain different additives, many of which are surface-active. They can assist each other, resulting in a synergistic effect, or they can have antagonistic effects. Many additives have several functions (multipurpose additives). These additives and base stocks are the elements used by the lubricant designer to meet the increasingly critical requirements of equipment manufacturers and users of lubricating oils and greases.

The following additives are blended with the base oil:

Detergents: These are surfactants to cleanse the harmful carbon and sludge deposits on the surface of the metals in contact. Sodium or calcium sulfonates or organic sulfonates are excellent detergent agents in lube oils.

Dispersants: These are used to disperse the oil-insoluble products of oxidation and other formations in the oil phase and do not allow these to deposit on the metallic surfaces of bearing rolling or sliding metals. Examples of dispersants are succinimides, esters of polysuccinic acid or succinate ester, and hydroxyethyl imide.

Antioxidants and stabilizers: These agents prevent auto-oxidation of hydrocarbon base oils present in the lubricant. This chemical reaction is in three stages: initiation, propagation, and termination, similar to a polymerization reaction forming resinous layers. Copper soaps are an excellent retardant of such auto-oxidation. Aromatic amines and phenols are examples of antioxidants.

Viscosity index improvers: Polymethacrylates and polyisobutylene are excellent viscosity index (VI) improvers. These agents keep the viscosity of oil nearly unchanged over a wide range of temperature fluctuations.

Friction modifiers: These agents modify the coefficient of friction by adhering to the metallic surface. Examples are amines, amides, their derivatives, carboxylic acids, phosphoric acids, and their salts.

Pour point depressants: Polymethacrylates, polyacrylates, and di-tetra paraffinphenol-phthalate act as the pour point suppressants.

Demulsifiers: Acidic gases, moisture, carbon particles, and other products may form at high temperatures, especially in engines, and form an emulsion with lubricant oils. Sulfonates, alkylated phenolic resins, polyethylene oxide, etc., are good demulsifiers.

Anti-foaming agents: Gases and moisture are responsible for foam formation with the lubricating oil. The most widely used anti-foaming agent is polydimethylsiloxane.

Corrosion inhibitors: Ingress of oxygen and the presence of moisture cause oxide corrosion, and acidic chemicals and mercaptans may cause chemical corrosion aided by high temperatures. Esters or amides of dodecyl-succinic acid, thiophosphates, etc., act as corrosion inhibitors. Anti-oxidants also prevent oxide corrosion.

Thickeners: Sodium or calcium soaps act as thickeners, which are required to retain the film of lubricant over the metallic surfaces in contact and do not allow the metallic surfaces to come in direct contact without the lubricant film within them.

A variety of lubricants is used depending on the type of application, such as automobiles, aircraft, ships, and engines. These are broadly classified as automotive lubricants, aviation lubricants, industrial lubricants, marine lubricants, etc. Under this broad classification, they are further classified as engine oil, gear oil, bearing oil, hydraulic oil or transmission oil, cylinder oil, etc., depending on the field of application.

1.8   Lube Blending

Various intermediate products, such as straight run and vacuum distillates, reformats, and desulfurized distillates, are received as rundown streams, many of which are later blended with other suitable streams to adjust the desired properties. For example, the motor spirit is a blend of a variety of streams, which are debutants reformate naphtha, FCC-cracked gasoline, thermally cracked gasoline, vis-broken naphtha, pyrolysis gasoline from petrochemical plants, etc. These are blended in proper proportions for adjusting the octane number, vapor pressure, oxidation stability, etc. During the winter season and for low-temperature climates, the motor spirit is also blended with butane to adjust the vapor pressure. Similarly, FO is a mixture of vis-broken vacuum distillates, asphalt, short residue, etc., and blending is carried out to correct the viscosity and the flash point.

The large number of natural lubricating oils sold today is produced by blending a small amount of lubricating oil base stocks and additives. The lube oil base stocks are prepared from selected crude oils by distillation and special processing to meet the desired qualifications. The additives are chemicals used to give the base stocks desirable characteristics that they lack or to enhance and improve existing properties.

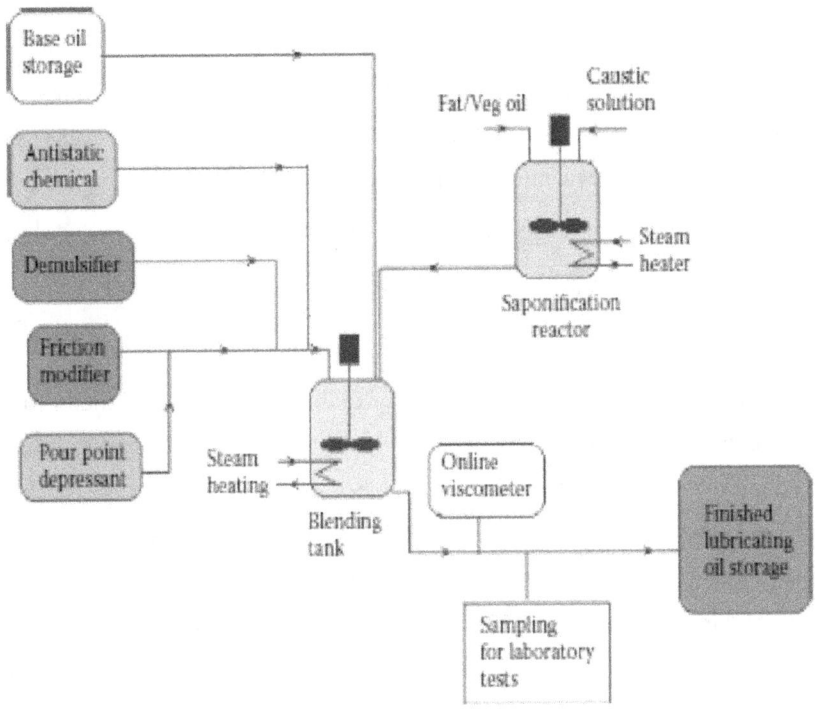

Figure 6: Schematic lube blending plant (Wiley, 2007)

Lube blending is the intelligent art of mixing various ingredients with the lube base stock, satisfying the desired physical, chemical, transport and mechanical properties, usually known as tribological properties. The blending units consist of kettles with steam or electric heating facilities where the additives are mixed while continuously monitoring the desired properties. Additives, such as anti-wear, antioxidant, anti-rust, anti-foam, and corrosion inhibitors, are given in small quantities and are procured from various sources. Usually lube blending and grease making are carried out in the same plant where large amounts of soaps of fatty oils are blended for grease and hence a saponification unit is part of a blending plant. A schematic general lube blending plant is shown in Figure 6.

CHAPTER 2

GREASES

Grease is thickened oil containing the base oil, thickening agent (gallant) and additives. The mineral base oil must meet many different requirements, e.g., viscosity, VI, and oxidation stability, such as that required for making lubricating oil. However, the other important properties for base oil required for grease making are the viscosity-gravity-constant (VGC), the aniline point, carbon type, and solubilising property. The thickeners are the sodium, calcium, or lithium soaps of fatty acids.

Greases are classified according to the type of thickeners used. The melting point and the water content of the soaps are important parameters. Sodium soaps (melt at 150°C) are preferred over calcium soaps (melt at 100°C) for their higher melting point. The evaporation of water present in these soaps limits their use much below their melting points. Lithium-12 hydroxystearate soap is an excellent thickener for grease making. This grease is able to maintain its properties at high temperatures without losing its lubricating property up to 140°C along with excellent resistance to mechanical deformation and immunity to the presence of water in the field of application. Modern grease employs a mixture of calcium– lithium and aluminium–barium soaps.

A major thickener in the grease industry is lithium 12-hydroxystearate, a soap of 12hydroxystearic acid, obtained from vegetable oils and animal fats. In place of metallic soaps, thickening agents, such as clays, PTFE powders, and graphite, are also used in grease. However, these require frequent replacement of grease to maintain uniform lubricity. Most modern greases also include some additives to modify the property of the grease to suit different environments of applications. These are antioxidants, corrosion inhibitors, antiwear, extreme pressure additives, adhesives, etc.

Grease behaves as non-Newtonian fluids and the majority usually manifest as Bingham plastics. The consistency index or apparent viscosity is measured for grease. Apparent viscosity is measured as per the ASTM D1092 method and consistency is measured by cone penetration index as per the ASTM D1403 method. A worked grease sample, obtained through a grease worker, which is a cylindrical vessel equipped with a piston with holes, is prepared before testing for cone penetration. Standard working is done by forcing the piston back and forth 60 times in 60 sec. The penetration index is measured at 25°C. In addition to these, the drop point is also determined according to the ASTM D556 method. Grease specification is classified according to the National Lubricating Grease Institute (NLGI). The properties and uses of typical greases are presented in Table 1.

Table 2: Typical specification for greases

Grade Number	Worked Penetration, mm^{-1}	Approx. Yield Strength, Pa	Approx. Self-supporting Height, cm (inches)	Description
000	445 to 475			Very fluid
00	400 to 430	90		Fluid
0	355 to 385	130	1.30 (0.5)	Semifluid
1	310 to 340	180	1.80 (0.7)	Very soft
2	265 to 295	300	3.00 (1.2)	Soft
3	220 to 250	560	5.60 (2.2)	Semifirm
4	175 to 205	1,300	13.0 (5.2)	Firm
5	130 to 160	3,800	38.0 (15.0)	Very firm
6	85 to 115			Hard

2.1 Manufacture of Grease

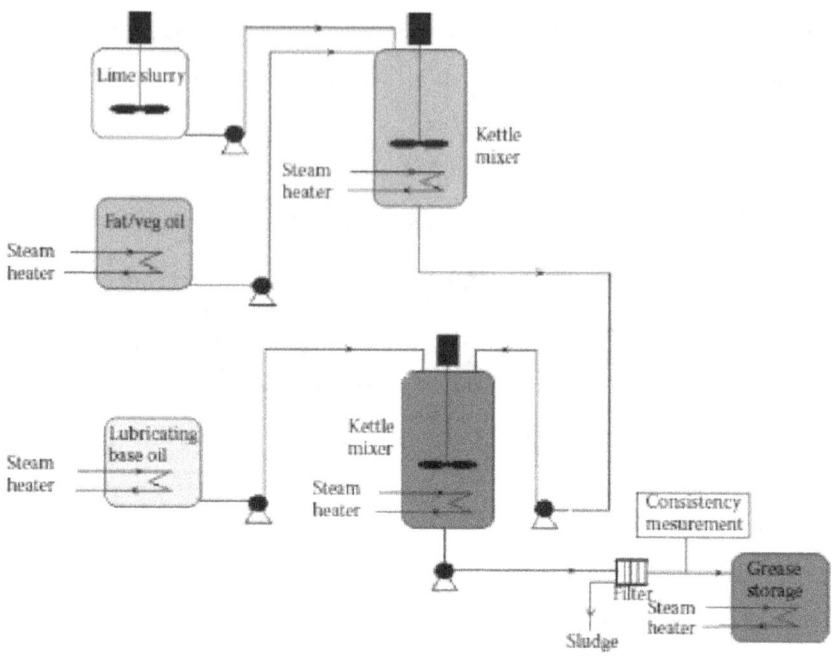

Figure 7: A calcium grease making plant (Chaudhuri, 2011).

A typical calcium grease-making plant is presented in Figure 7, where calcium soap is prepared from metered amounts of lime slurry and vegetable oil at the desirable operating temperature and oil/lime ratio in a continuously stirred steam-heated tank or a kettle heater. Modern plants use electric heating in the kettles. Several kettles may be used to increase the production rate. Fats or vegetable oil are saponified and a soap mixture is pumped to a special type of kettle, known as a Lancaster mixer, where the base oil (usually petroleum base) is mixed at a high speed of about 3000–4000 rpm and a temperature of 170°C–180°C. The operating temperature and time of mixing are monitored for the desired consistency of the grease. The effluent from the kettle is then filtered to remove sludge and particulates from the liquid grease before storage or continuous packaging.

CHAPTER 3

BITUMEN MANUFACTURE

3.1   What is Bitumen?

Bitumen (also known as native asphalt or extra heavy oil) is a common binder that is used for the construction of roads. It exists in nature as reddish brown or black materials of semisolid, viscous to brittle character with no mineral impurity or with mineral matter contents, that exceed 50% by weight. It also is obtained as a residual product in the refining of petroleum. Bitumen is a naturally occurring material that is found in deposits filling the pores and crevices of sandstone, limestone, or argillaceous sediments where the permeability is low and passage of fluids through the deposit can only be achieved by prior application of fracturing techniques. The recovery of the bitumen depends to a large degree on the composition and construction of the sands. The bitumen in tar sand formations requires a high degree of thermal stimulation for recovery to the extent that some thermal decomposition may have to be induced. Currently, recovery operations of bitumen in tar sand formations involve the use of mining techniques. Recovery methods are based either on mining combined with some further processing or operation on the oil sands in situ. The mining methods apply to shallow deposits, characterized by an overburden ratio (i.e., overburden depth to the thickness of tar-sand deposit). For example, indications are that for the Athabasca deposit, no more than 15% of the in-place deposit is available within current concepts of the economics and technology of open-pit mining; this 10% portion may be considered as the proven reserves of bitumen in the deposit.

3.2   Characteristics of Bitumen

   a. Bitumen is colloidal and thermoplastic
   b. Bitumen has a low melting point
   c. It is insoluble in water
   d. It is impermeable to water (hydrophobic)
   e. It oxidizes slowly

These characteristics and the following, are what make bitumen a good construction material:

   i   Production of bitumen is economical
   ii It can be recycled by asphalt recycling process

iii It is not toxic

iv Can form several colors with the addition of pigment

3.3 Refinery Units that Produce Bitumen

Bitumen or Asphalt is obtained from short residue (residual mass from the bottom of the vacuum distillation unit) after propane extraction of valuable oil (known as bright stock). Bitumen is obtained from the heavier fraction because, it has more asphaltene, molecular weight, and higher penetration grade.

The vacuum distillation unit vacuum distillation unit (VDU) produces hard and soft bitumen. If the vacuum bottom is soft bitumen, it can be deasphalted; removing oil in a liquid-liquid extraction at low temperature but high pressure. If the vacuum bottom is an intermediate bitumen, it can be blown by air in a blowing tower to articulate molecules.

3.4 Bitumen Blowing

Raw asphalt from the deasphalting unit is blown with hot air in a furnace to adjust the softening point and penetration index for the production of paving-grade bitumen. Depending on the surface temperature and the environment of the application, the softening temperature and penetration index are adjusted by varying the air/feed ratio, temperature, and blowing time in the furnace. A typical bitumen-blowing unit is shown in Figure 8. If the asphalt contains lower amounts of metals, these can be routed to the coking unit for the production of metallurgical coke. Lube-bearing crude oil yields asphalt of high metal content whereas non-lube-bearing crudes yield asphalts with low metallic contents. As a result, asphalts from lube-bearing crude are suitable for the production of bitumen, whereas asphalts, SRs, and even heavy vacuum distillates from non-lube-bearing crude are suitable for coke production.

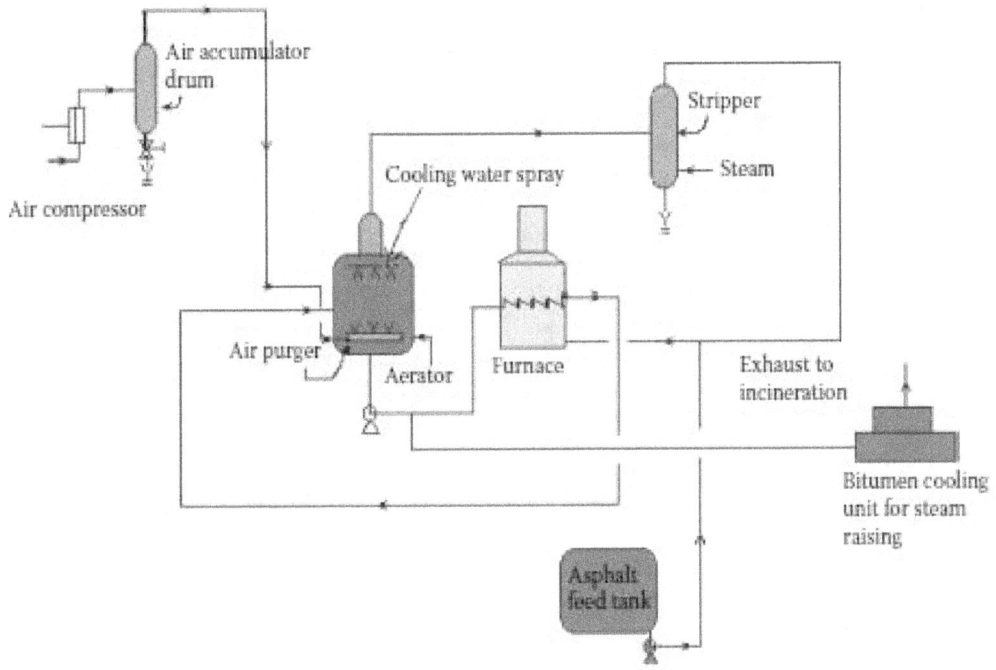

Figure 8: A typical bitumen-blowing unit for the production of paving grade bitumen (Chaudhuri, 2011).

3.5   Properties of Paving Grade Bitumen

Bitumen have desirable penetration index, which is defined as the depth of submergence or penetration of a standard weight through a needle penetrator. The greater the penetration, the greater the softness of the asphalt. Typical paving grade bitumen has a penetration of 60/70 or 80/100 using a 100-g cone (1/100 cm) at 25°C for 5 sec during the test. The other property is the flash point. As this has to be applied on the open space surface, the flash point should be above 175°C and the softening point should be above the ambient temperature depending on requirements. In addition to these, mechanical properties like ductility must also be measured. A list of the important properties of a typical paving grade bitumen cited in (Chaudhuri, 2011) is given below.

Typical Specifications of Bitumen

Property	Specification	
Penetration index	60/70	80/100
Flash point (PMC), min	175°C	175°C
Softening point, min	40-55°C	35-50°C
Matter soluble in carbon disulfide, min	99% wt	99% wt
Density at 15°C	0.99	0.99
Ductility at 27°C in centimetres, min	75	75

You can also study the specifications of Bitumen shown below:

Font Size	Standard Test	Bitumen (asphalt type)	
		AC 40/50	AC 60/70
Density	ASTM D7	1.01-1.06	1.01-1.06
Penetration @25°C, 10/mm	ASTM D5	40-50	60-70
Softening Point (°C)	ASTM D36	52-60	49-56
Ductility at 25°C (cm)	ASTM D113	Min 100	Min 100
Flash Point (°C)	ASTM D92	Min 250	Min 250
Solubility in Disulfide % wt	ASTM D4	Min 99.5	Min 99.5
Strain Test	AASHTO 102	Negative	Negative
Weight Loss by Heating % wt	ASTM D6	Max 0.2	Max 0.2
Penetration Loss by Heating %	ASTM D5	Max 20	Max 20

Bitumen blending is sometimes necessary to correct the penetration index, flash, and softening points. Internal fuel oil (IFO) for consumption within a refinery is a mixture of asphalt, residue, wax, etc. Proper blending may be necessary for these components to be used in the furnaces of process units and the power plant.

3.6   Types of Bitumen

The following types of bitumen are produced from VDU bottoms:

- ➤ Penetration grade bitumen
- ➤ Straight run bitumen
- ➤ Oxidized bitumen
- ➤ Polymer modified bitumen
- ➤ Cut-back bitumen
- ➤ Bitumen emulsion

CHAPTER 4

CRUDE OIL DISTILLATION, THE DEFINITION OF CAPACITY REFINERY AND OTHER PROCESSES

A crude oil refinery is a group of industrial facilities that turns crude oil and other inputs into finished petroleum products. A refinery's capacity refers to the maximum amount of crude oil designed to flow into the distillation unit of a refinery, also known as the crude unit.

The diagram below presents a stylized version of the distillation process. Crude oil is made up of a mixture of hydrocarbons, and the distillation process aims to separate this crude oil into broad categories of its component hydrocarbons, or "fractions." Crude oil is first heated and then put into a distillation column, also known as a still, where different products boil off and are recovered at different temperatures.

Lighter products, such as butane and other liquid petroleum gases (LPG), gasoline blending components, and naphtha, are recovered at the lowest temperatures. Mid-range products include jet fuel, kerosene, and distillates (such as home heating oil and diesel fuel). The heaviest products such as residual fuel oil are recovered at temperatures sometimes over 1,000 degrees Fahrenheit.

The simplest refineries stop at this point. Although not shown in the simplified diagram above, most refineries in the United States reprocess the heavier fractions into lighter products to maximize the output of the most desirable products using more sophisticated refining equipment such as catalytic crackers, reformers, and cokers.

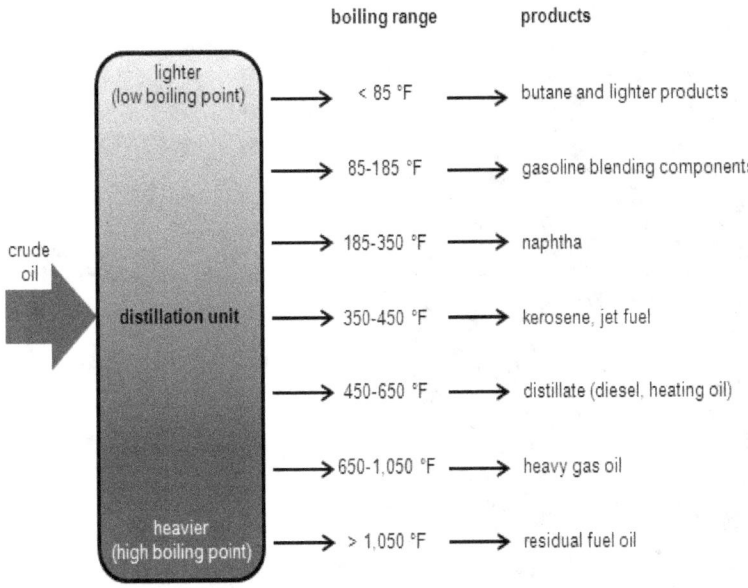

boiling range   products

lighter (low boiling point) → < 85 °F → butane and lighter products

→ 85-185 °F → gasoline blending components

→ 185-350 °F → naphtha

crude oil → distillation unit → 350-450 °F → kerosene, jet fuel

→ 450-650 °F → distillate (diesel, heating oil)

→ 650-1,050 °F → heavy gas oil

heavier (high boiling point) → > 1,050 °F → residual fuel oil

Fig 4.1   Crude oil distillation unit and product

4.1   VACUUM DISTILLATION

The principles of vacuum distillation resemble those of fractional distillation (commonly called atmospheric distillation to distinguish it from the vacuum method), except that larger-diameter columns are used to maintain comparable vapor velocities at reduced operating pressures. A vacuum of 50 to 100 mm of mercury absolute is produced by a vacuum pump or steam ejector.

The primary advantage of vacuum distillation is that it allows for distilling heavier materials at lower temperatures than those that would be required at atmospheric pressure, thus avoiding thermal cracking of the components. Firing conditions in the furnace are adjusted so that oil temperatures usually do not exceed 425 °C (800 °F). The residue remaining after vacuum distillation, called bitumen, may be further blended to produce road asphalt or residual fuel oil, or it may be used as a feedstock for thermal cracking or coking units. Vacuum distillation units are essential parts of the many processing schemes designed to produce lubricants.

Fig 4.1 (a) Vacuum distillation unit in the refinery

The schematic of the vacuum distillation unit is shown below for a better understanding.

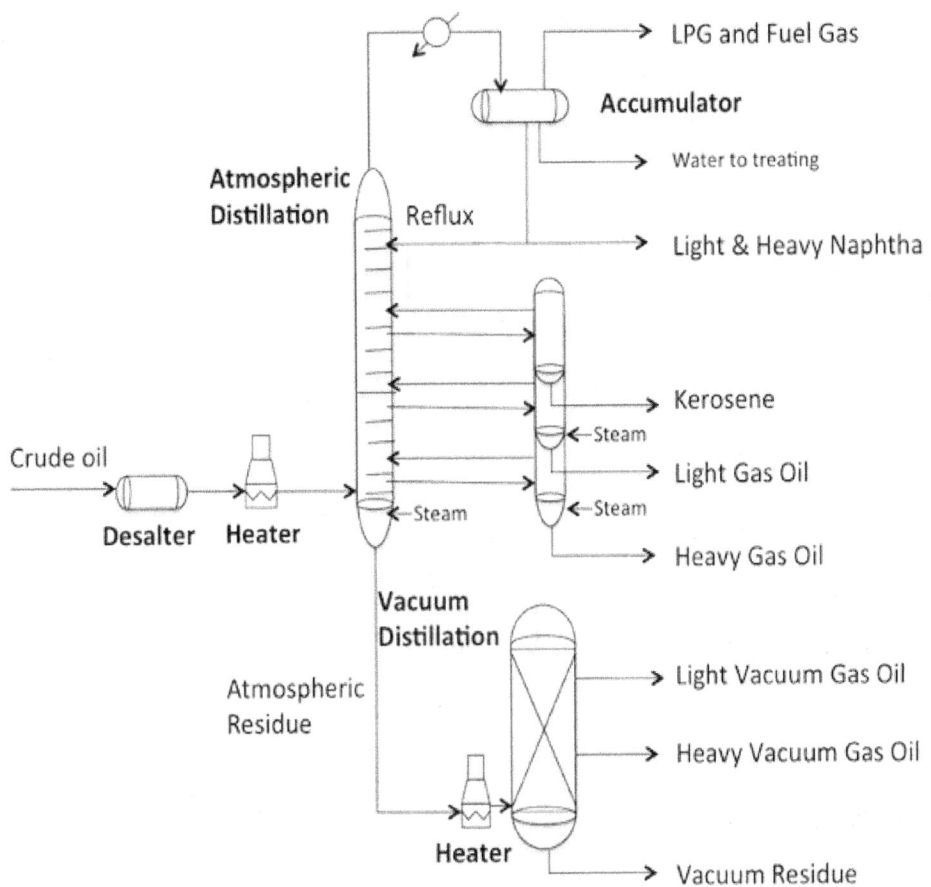

Fig 4.1 (b) Atmospheric and Vacuum distillation unit

5.2   SUPERFRACTIONATION

An extension of the distillation process, super-fractionation employs smaller-diameter columns with a much larger number of trays (100 or more) and reflux ratios exceeding 5:1. With such equipment it is possible to isolate a very narrow range of components or even pure compounds. Common applications involve the separation of high-purity solvents such as isoparaffins or individual aromatic compounds for use as petrochemicals.

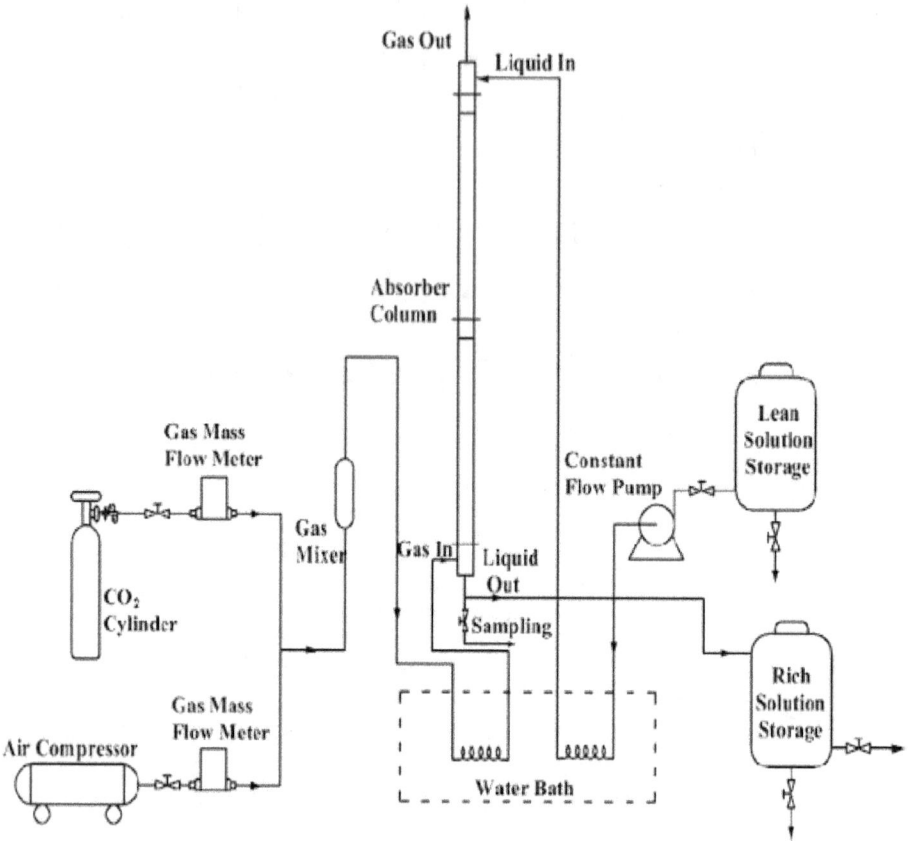

5.3 ABSORPTION

Absorption processes are employed to recover valuable light components such as propane/propylene and butane/butylene from the vapors that leave the top of crude oil or process-unit fractionating columns within the refinery. These volatile gases are bubbled through an absorption fluid, such as kerosene or heavy naphtha, in equipment resembling a fractionating column. The light products dissolve in the oil while the dry gases—such as hydrogen, methane, ethane, and ethylene—pass through undissolved. Absorption is more effective under pressures of about 7 to 10 bars (0.7 to 1 megapascal [MPa]), or 100 to 150 psi, than it is at atmospheric pressure.

The enriched absorption fluid is heated and passed into a stripping column, where the light product vapors pass upward and are condensed for recovery as liquefied petroleum gas

(LPG). The unvaporized absorption fluid passes from the base of the stripping column and is reused in the absorption tower.

4.4   ADSORPTION

Certain highly porous solid materials can select and adsorb specific types of molecules, thus separating them from other materials. Silica gel is used in this way to separate aromatics from other hydrocarbons, and activated charcoal is used to remove liquid components from gases. Adsorption is thus somewhat analogous to the process of absorption with an oil, although the principles are different. Layers of adsorbed material only a few molecules thick are formed on the extensive interior surface of the adsorbent; the interior surface may amount to several hectares per kilogram of material.

Molecular sieves are a special form of adsorbent. Such sieves are produced by the dehydration of naturally occurring or synthetic zeolites (crystalline alkali-metal aluminosilicates). The dehydration leaves intercrystalline cavities that have pore openings of definite size, depending on the alkali metal of the zeolite. Under adsorptive conditions, normal paraffin molecules can enter the crystalline lattice and be selectively retained, whereas all other molecules are excluded. This principle is used commercially for the removal of normal paraffins from gasoline fuels, thus improving their combustion properties. The use of molecular sieves is also extensive in the manufacture of high-purity solvents.

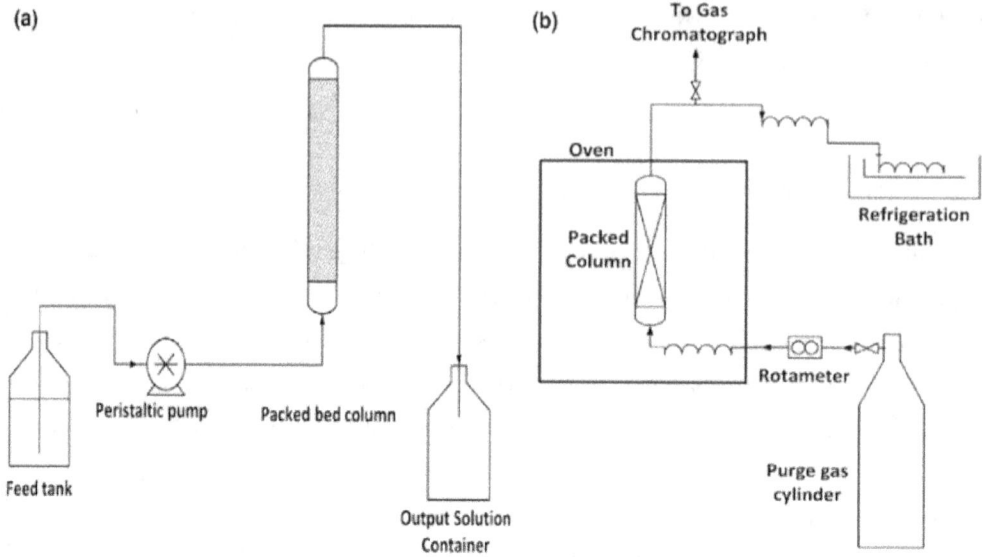

(a)

Feed tank

Peristaltic pump   Packed bed column

Output Solution
Container

(b)

To Gas
Chromatograph

Oven

Packed
Column

Refrigeration
Bath

Rotameter

Purge gas
cylinder

Fig 4.4 Schematic diagram of Adsorption

4.5   CRYSTALLIZATION

The crystallization of wax from lubricating oil fractions is essential to make oils suitable for use. A solvent (often a mixture of benzene and methyl ethyl ketone) is first added to the oil, and the solution is chilled to about −20 °C (−5 °F). The function of the benzene is to keep the oil in solution and maintain its fluidity at low temperatures, whereas the methyl ethyl ketone acts as a wax precipitant. Rotary filters deposit the wax crystals on a specially woven cloth stretched over a perforated cylindrical drum. A vacuum is maintained within the drum to draw the oil through the perforations. The wax crystals are removed from the cloth by metal scrapers, after washing with solvent to remove traces of oil. The solvents are later distilled from the oil and reused.

4.6   CONVERSION

The separation processes described above are based on differences in the physical properties of the components of crude oil. All petroleum refineries throughout the world employ at least crude oil distillation units to separate naturally occurring fractions for further use, but those that employ distillation alone are limited in their yield of valuable transportation fuels. By adding more complex conversion processes that chemically change the molecular structure of naturally occurring components of crude oil, it is possible to increase the yield of valuable hydrocarbon compounds.

4.7   Naphtha reforming

The most widespread process for rearranging hydrocarbon molecules is naphtha reforming. The initial process, thermal reforming, was developed in the late 1920s. Thermal reforming employed temperatures of 510–565 °C (950–1,050 °F) at moderate pressures—about 40 bars (4 MPa), or 600 psi—to obtain gasoline (petrol) with octane numbers of 70 to 80 from heavy naphthas with octane numbers of less than 40. The product yield, although of a higher octane level, included olefins, diolefins, and aromatic compounds. It was therefore inherently unstable in storage and tended to form heavy polymers and gums, which caused combustion problems.

By 1950 a reforming process was introduced that employed a catalyst to improve the yield of the most desirable gasoline components while minimizing the formation of unwanted heavy products and coke. (A catalyst is a substance that promotes a chemical reaction but does not take part in it.) In catalytic reforming, as in thermal reforming, a naphtha-type material serves as the feedstock, but the reactions are carried out in the presence of hydrogen, which inhibits the formation of unstable unsaturated compounds that polymerize into higher-boiling materials.

In most catalytic reforming processes, platinum is the active catalyst; it is distributed on the surface of an aluminum oxide carrier. Small amounts of rhenium, chlorine, and fluorine act as catalyst promoters. Despite the high cost of platinum, the process is economical because of the long life of the catalyst and the high quality and yield of the products obtained. The principal reactions involve the breaking down of long-chain hydrocarbons into smaller saturated chains and the formation of isoparaffins, made up of branched-chain molecules.

48

Formation of ring compounds (technically, the cyclization of paraffins into naphthenes) also takes place, and the naphthenes are then dehydrogenated into aromatic compounds (ring-shaped unsaturated compounds with fewer hydrogen atoms bonded to the carbon). The hydrogen liberated in this process forms a valuable by-product of catalytic reforming. The desirable end products are isoparaffins and aromatics, both having high octane numbers.

In a typical reforming unit the naphtha charge is first passed over a catalyst bed in the presence of hydrogen to remove any sulfur impurities. The desulfurized feed is then mixed with hydrogen (about five molecules of hydrogen to one of hydrocarbon) and heated to a temperature of 500–540 °C (930–1,000 °F). The gaseous mixture passes downward through catalyst pellets in a series of three or more reactor vessels. Early reactors were designed to operate at about 25 bars (2.5 MPa), or 350 psi, but current units frequently operate at less than 7 bars (0.7 MPa), or 100 psi. Because heat is absorbed in reforming reactions, the mixture must be reheated in intermediate furnaces between the reactors.

Block Diagram of Reformer

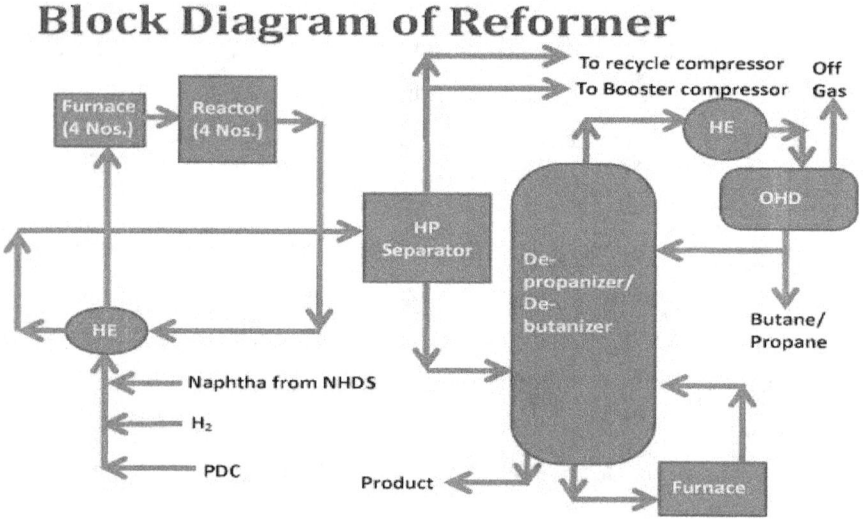

After leaving the final reactor, the product is condensed to a liquid, separated from the hydrogen stream, and passed to a fractionating column, where the light hydrocarbons produced in the reactors are removed by distillation. The reformated product is then available for blending into gasoline without further treatment. The hydrogen leaving the product separator is compressed and returned to the reactor system.

Operating conditions are set to obtain the required octane level, usually between 90 and 100. At the higher octane levels, product yields are smaller, and more frequent catalyst regenerations are required. During the reforming process, minute amounts of carbon are deposited on the catalyst, causing a gradual deterioration of the product yield pattern. Some units are semi-regenerative facilities—that is, they must be removed from service periodically (once or twice annually) to burn off the carbon and rejuvenate the catalyst system—but increased demand for high-octane fuels has also led to the development of continuous regeneration systems, which avoid the periodic unit shutdowns and maximize the yield of high-octane reformate. Continuous regeneration employs a moving bed of catalyst particles that is gradually withdrawn from the reactor system and passed through a regenerator vessel, where the carbon is removed and the catalyst rejuvenated for reintroduction to the reactor system.

4.8   Catalytic cracking

The use of thermal cracking units to convert gas oils into naphtha dates from before 1920. These units produced small quantities of unstable naphthas and large amounts of by-product coke. While they succeeded in providing a small increase in gasoline yields, it was the commercialization of the fluid catalytic cracking process in 1942 that established the foundation of modern petroleum refining. The process not only provided a highly efficient means of converting high-boiling gas oils into naphtha to meet the rising demand for high-octane gasoline, but it also represented a breakthrough in catalyst technology.

The thermal cracking process functioned largely by the free-radical theory of molecular transformation. Under conditions of extreme heat, the electron bond between carbon atoms in a hydrocarbon molecule can be broken, thus generating a hydrocarbon group with an unpaired electron. This negatively charged molecule, called a free radical, enters into reactions with other hydrocarbons, continually producing other free radicals via the transfer of negatively charged hydride ions (H−). Thus a chain reaction is established that leads to a reduction in molecular size, or "cracking," of components of the original feedstock.

The use of a catalyst in the cracking reaction increases the yield of high-quality products under much less severe operating conditions than in thermal cracking. Several complex reactions are involved, but the principal mechanism by which long-chain hydrocarbons are cracked into lighter products can be explained by the carbonium ion theory. According to this theory, a catalyst promotes the removal of a negatively charged hydride ion from a paraffin compound or the addition of a positively charged proton ($H+$) to an olefin compound. This results In the formation of a carbonium ion, a positively charged molecule that has only a very short life as an intermediate compound that transfers the positive charge through the hydrocarbon. Carbonium transfer continues as hydrocarbon compounds come into contact with active sites on the surface of the catalyst that promote the continued addition of protons or removal of hydride ions. The result is a weakening of carbon-carbon bonds in many of the hydrocarbon molecules and a consequent cracking into smaller compounds.

Olefins crack more readily than paraffins since their double carbon-carbon bonds are more friable under reaction conditions. Isoparaffins and naphthenes crack more readily than normal paraffins, which in turn crack faster than aromatics. Aromatic ring compounds are very resistant to cracking since they readily deactivate fluid-cracking catalysts by blocking the active sites of the catalyst.

Principal reactions in fluid catalytic cracking

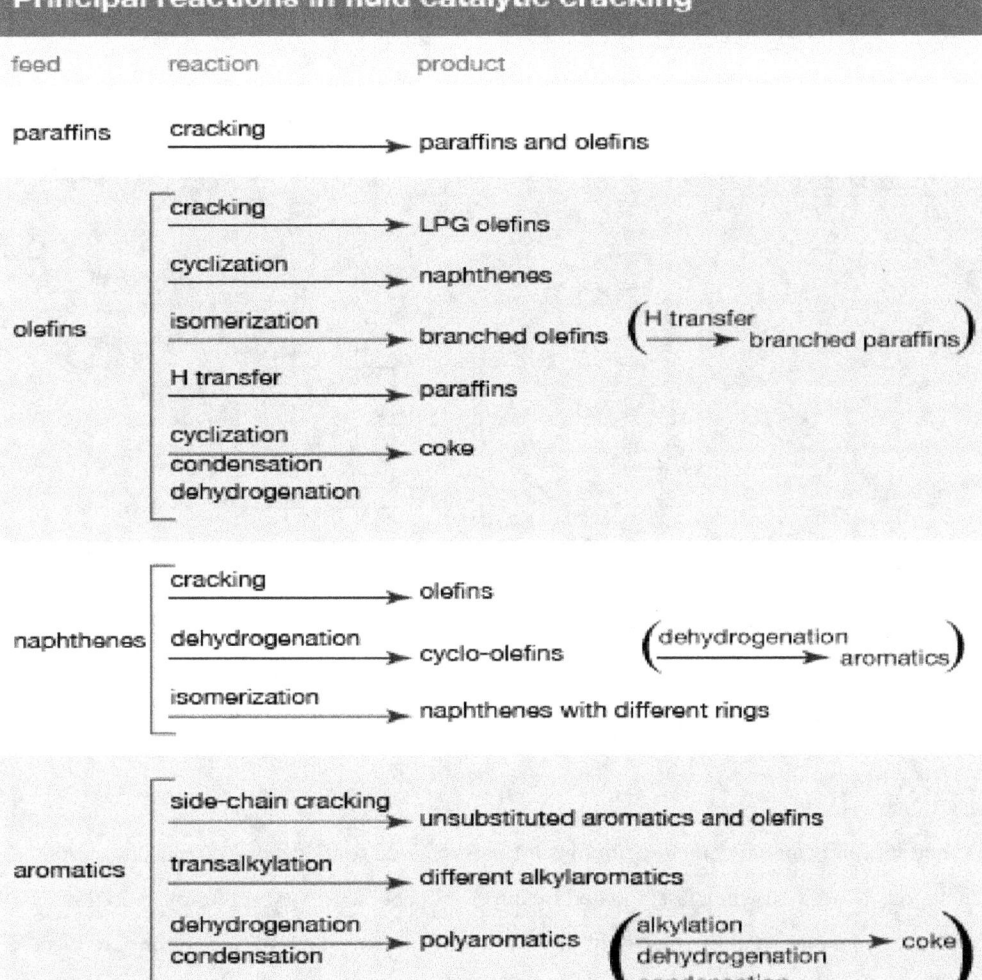

feed	reaction	product
paraffins	cracking	paraffins and olefins

olefins
- cracking → LPG olefins
- cyclization → naphthenes
- isomerization → branched olefins (H transfer → branched paraffins)
- H transfer → paraffins
- cyclization / condensation / dehydrogenation → coke

naphthenes
- cracking → olefins
- dehydrogenation → cyclo-olefins (dehydrogenation → aromatics)
- isomerization → naphthenes with different rings

aromatics
- side-chain cracking → unsubstituted aromatics and olefins
- transalkylation → different alkylaromatics
- dehydrogenation / condensation → polyaromatics (alkylation / dehydrogenation / condensation → coke)

The table illustrates many of the principal reactions that are believed to occur in fluid catalytic cracking unit reactors. The reactions postulated for olefin compounds apply principally to intermediate products within the reactor system since the olefin content of the catalytic cracking feedstock is usually very low.

Typical modern catalytic cracking reactors operate at 480–550 °C (900–1,020 °F) and at relatively low pressures of 0.7 to 1.4 bars (70 to 140 KPa), or 10 to 20 psi. At first natural silica-alumina clays were used as catalysts, but by the mid-1970s zeolitic and molecular sieve-based catalysts became common. Zeolitic catalysts give more selective yields of products while reducing the formation of gas and coke.

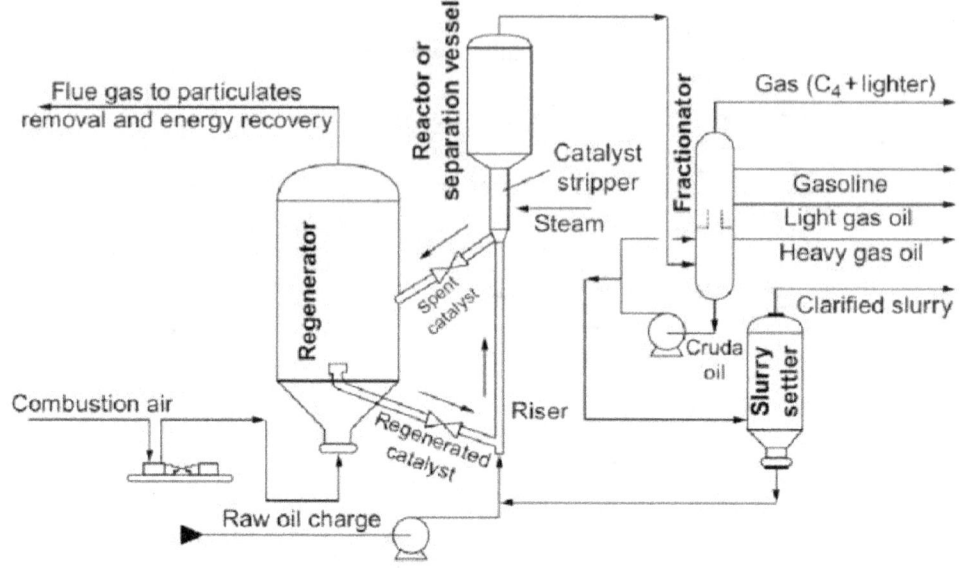

Fig 4.8: Detailed schematic diagram for fluid catalytic cracking unit

A modern fluid catalytic cracker employs a finely divided solid catalyst that has properties analogous to a liquid when it is agitated by air or oil vapors. The principles of operation of such a unit are shown in the figure. In this arrangement, a reactor and regenerator are located side by side. The oil feed is vaporized when it meets the hot catalyst at the feed-injection point, and the vapors flow upward through the riser reactor at high velocity, providing a fluidizing effect for the catalyst particles. The catalytic reaction occurs exclusively in the riser reactor. The catalyst then passes into the cyclone vessel, where it is separated from reactor hydrocarbon products.

As the cracking reactions proceed, carbon is deposited on the catalyst particles. Since these deposits impair the reaction efficiency, the catalyst must be continuously withdrawn from the reaction system. Unit product vapors pass out of the top of the reactor through cyclone separators, but the catalyst is removed by centrifugal force and dropped back into the stripper section. In the stripping section, hydrocarbons are removed from the spent catalyst with steam, and the catalyst is transferred through the stripper standpipe to the regenerator vessel, where the carbon is burned with a current of air. The high temperature of the regeneration process (675–785 °C, or 1,250–1,450 °F) heats the catalyst to the desired reaction temperature for recontacting fresh feed into the unit. To maintain activity, a small amount of fresh catalyst is added to the system from time to time, and a similar amount is withdrawn.

The cracked reactor effluent is fractionated in a distillation column. The yield of light products (with boiling points less than 220 °C, or 430 °F) is usually reported as the conversion level for the unit. Conversion levels average about 60 to 70 percent in Europe and Asia and over 80 percent in many catalytic cracking units in the United States. About one-third of the product yield consists of fuel gas and other gaseous hydrocarbons. Half of this is usually propylene and butylene, which are important feedstocks for the polymerization and alkylation processes discussed below. The largest volume is usually cracked naphtha, an important gasoline blend stock with an octane number of 90 to 94. The lower conversion units of Europe and Asia produce comparatively more distillate oil and less naphtha and light hydrocarbons.

4.9   POLYMERIZATION AND ALKYLATION

The light gaseous hydrocarbons produced by catalytic cracking are highly unsaturated and are usually converted into high-octane gasoline components in polymerization or alkylation processes. In polymerization, the light olefins propylene and butylene are induced to combine, or polymerize, into molecules of two or three times their original molecular weight. The catalysts employed consist of phosphoric acid on pellets of kieselguhr, a porous sedimentary rock. High pressures, on the order of 30 to 75 bars (3 to 7.5 MPa), or 400 to 1,100 psi, are required at temperatures ranging from 175 to 230 °C (350 to 450 °F). Polymer gasoline derived from propylene and butylene has octane numbers above 90.

The alkylation reaction also achieves a longer chain molecule by the combination of two smaller molecules, one being an olefin and the other an isoparaffin (usually isobutane). During World War II, alkylation became the main process for the manufacture of isooctane, a primary component in the blending of aviation gasoline.

Two alkylation processes employed in the industry are based upon different acid systems as catalysts. In sulfuric acid alkylation, concentrated sulfuric acid of 98 percent purity catalyzes a reaction that is carried out at 2 to 7 °C (35 to 45 °F). Refrigeration is necessary because of the heat generated by the reaction. The octane numbers of the alkylates produced range from 85 to 95.

Hydrofluoric acid is also used as a catalyst for many alkylation units. The chemical reactions are similar to those in the sulfuric acid process, but it is possible to use higher temperatures (between 24 and 46 °C, or 75 to 115 °F), thus avoiding the need for refrigeration. Recovery of hydrofluoric acid is accomplished by distillation. Stringent safety precautions must be exercised when using this highly corrosive and toxic substance.

4.10   Hydrocracking

One of the most far-reaching developments of the refining industry in the 1950s was the use of hydrogen, made possible in part by the availability of hydrogen as a by-product of catalytic reforming. Since the 1980s hydrogen processing has become so prominent that many refineries now incorporate hydrogen-manufacturing plants in their processing schemes.

Though hydrocracking processes a similar feedstock to the catalytic cracking unit, it offers even greater flexibility in product yields. The process can be used for producing gasoline or jet fuels from heavy gas oils, for producing high-quality lubricating oils, or for converting distillation residues into lighter oils. The jet fuel and distillate oil products are of high quality and low sulfur content and may be blended into final products without further processing. Hydrocracked naphtha, on the other hand, is often low in octane and must be catalytically reformed to produce high-quality gasoline.

Hydrocracking is accomplished at lower temperatures than catalytic cracking—e.g., 260 to 425 °C (500 to 800 °F)—but at much higher pressures—55 to 170 bars (5.5 to 17 MPa), or 800 to 2,500 psi. The design and manufacture of large, thick-walled vessels for operation under these conditions has been a major engineering achievement.

Hydrocracking catalysts vary widely. The cracking reactions are induced by materials of the silica-alumina type. In units that process residual feedstocks, hydrogenation catalysts such as nickel, tungsten, platinum, or palladium are employed. The activity of the catalyst system can be maintained for long periods so that continuous regeneration is not necessary as in catalytic cracking.

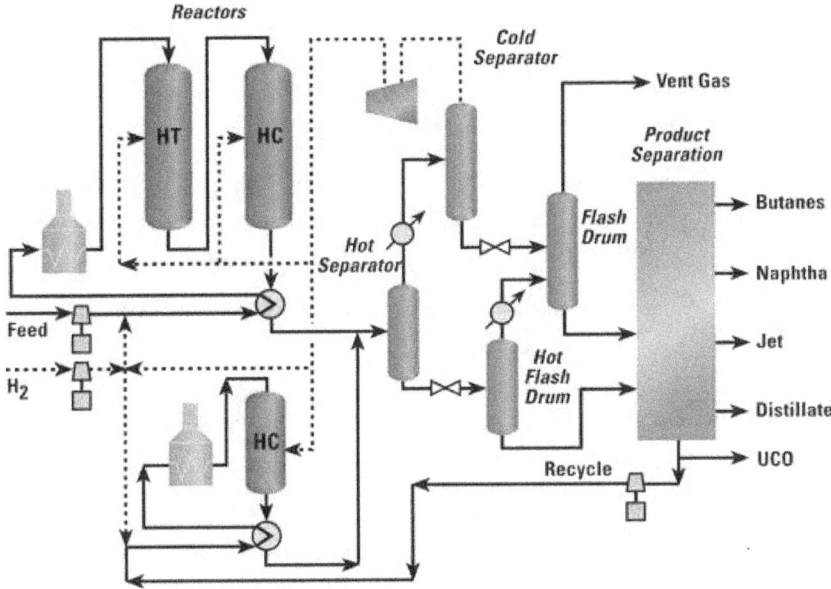

Fig 4.10: Hydrocracking process

4.11 ISOMERIZATION

The demand for aviation gasoline became so great during World War II and afterward that the quantities of isobutane available for alkylation feedstock were insufficient. This deficiency was remedied by isomerization of the more abundant normal butane into isobutane. The isomerization catalyst is aluminum chloride supported on alumina and promoted by hydrogen chloride gas.

Commercial processes have also been developed for the isomerization of low-octane normal pentane and normal hexane to the higher-octane isoparaffin form. Here the catalyst is usually promoted with platinum. As in catalytic reforming, the reactions are carried out in the presence of hydrogen. Hydrogen is neither produced nor consumed in the process but is employed to inhibit undesirable side reactions. The reactor step is usually followed by molecular sieve extraction and distillation. Though this process is an attractive way to exclude low-octane components from the gasoline blending pool, it does not produce a final product of sufficiently high octane to contribute much to the manufacture of unleaded gasoline.

4.12 VISBREAKING, THERMAL CRACKING, AND COKING

Since World War II the demand for light products (e.g., gasoline, jet, and diesel fuels) has grown, while the requirement for heavy industrial fuel oils has declined. Furthermore, many of the new sources of crude petroleum (California, Alaska, Venezuela, and Mexico) have yielded heavier crude oils with higher natural yields of residual fuels. As a result, refiners have become even more dependent on the conversion of residue components into lighter oils that can serve as feedstock for catalytic cracking units.

As early as 1920, large volumes of residue were being processed in visbreakers or thermal cracking units. These simple process units consist of a large furnace that heats the feedstock to the range of 450 to 500 °C (840 to 930 °F) at an operating pressure of about 10 bars (1 MPa), or about 150 psi. The residence time in the furnace is carefully limited to prevent much of the reaction from taking place and clogging the furnace tubes. The heated feed is then charged to a reaction chamber, which is kept at a pressure high enough to permit cracking of the large molecules but restricts coke formation. From the reaction chamber, the process fluid is cooled to inhibit further cracking and then charged to a distillation column for separation into components.

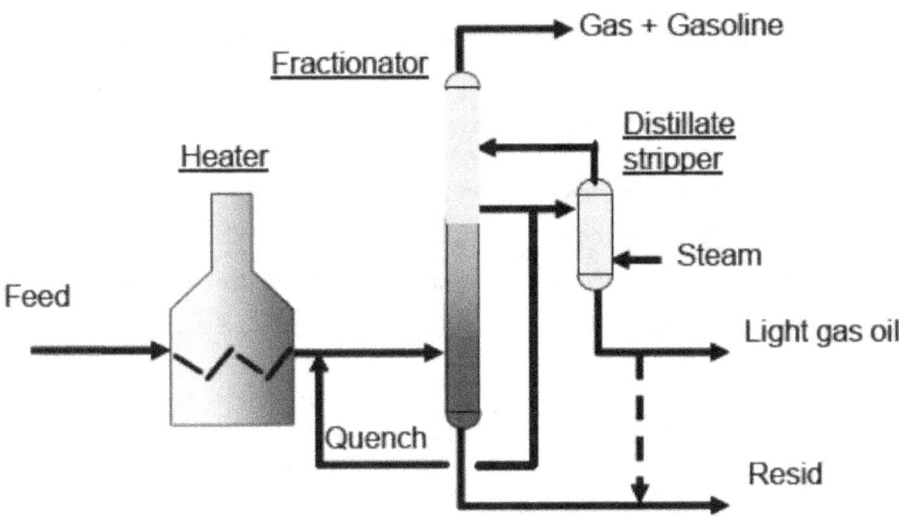

Fig 4.12: Diagram for the visbreaking process

Visbreaking units typically convert about 15 percent of the feedstock to naphtha and diesel oils and produce a lower-viscosity residual fuel. Thermal cracking units provide more severe processing and often convert as much as 50 to 60 percent of the incoming feed to naphtha and light diesel oils.

Coking is severe thermal cracking. The residue feed is heated to about 475 to 520 °C (890 to 970 °F) in a furnace with very low residence time and is discharged into the bottom of a large vessel called a coke drum for extensive and controlled cracking. The cracked lighter product rises to the top of the drum and is drawn off. It is then charged to the product fractionator for separation into naphtha, diesel oils, and heavy gas oils for further processing in the catalytic cracking unit. The heavier product remains and, because of the retained heat, cracks ultimately to coke, a solid carbonaceous substance akin to coal. Once the coke drum is filled with solid coke, it is removed from service and replaced by another coke drum.

Decoking is a routine daily occurrence accomplished by a high-pressure water jet. First, the top and bottom heads of the coke drum are removed. Next, a hole is drilled in the coke from the top to the bottom of the vessel. Then a rotating stem is lowered through the hole, spraying a water jet sideways. The high-pressure jet cuts the coke into lumps, which fall out the

58

bottom of the drum for subsequent loading into trucks or railcars for shipment to customers. Typically, coke drums operate on 24-hour cycles, filling with coke over one 24-hour period followed by cooling, decoking, and reheating over the next 24 hours. The drilling derricks on top of the coke drums are a notable feature of the refinery skyline.

Cokers produce no liquid residue but yield up to 30 percent coke by weight. Much of the low-sulfur product is employed to produce electrodes for the electrolytic smelting of aluminum. Most lower-quality coke is burned as fuel in admixture with coal. Coker economics usually favor the conversion of residue into light products even if there is no market for the coke.

4.13   PURIFICATION

Before petroleum products can be marketed, certain impurities must be removed or made less obnoxious. The most common impurities are sulfur compounds such as hydrogen sulfide (H2S) or the mercaptans ("R" SH)—the latter being a series of complex organic compounds having as many as six carbon atoms in the hydrocarbon radical ("R"). Apart from their foul odor, sulfur compounds are technically undesirable. In motor and aviation gasoline they reduce the effectiveness of antiknock additives and interfere with the operation of exhaust-treatment systems. In diesel fuel, they cause engine corrosion and complicate exhaust treatment systems. Also, many major residual and industrial fuel consumers are located in developed areas and are subject to restrictions on sulfurous emissions.

Most crude oils contain small amounts of hydrogen sulfide, but these levels may be increased by the decomposition of heavier sulfur compounds (such as the mercaptans) during refinery processing. The bulk of the hydrogen sulfide is contained in process-unit overhead gases, which are ultimately consumed in the refinery fuel system. To minimize noxious emissions, most refinery fuel gases are desulfurized.

Other undesirable components include nitrogen compounds, which poison catalyst systems, and oxygenated compounds, which can lead to color formation and product instability. The principal treatment processes are outlined below.

4.14   SWEETENING

Sweetening processes oxidize mercaptans into more innocuous disulfides, which remain in the product fuels. Catalysts assist in the oxidation. The doctor process employs sodium plumbite, a solution of lead oxide in caustic soda, as a catalyst. At one time this inexpensive process was widely practiced, but the necessity of adding elemental sulfur to make the reactions proceed caused an increase in total sulfur content in the product. It has largely been replaced by the copper chloride process, in which the catalyst is a slurry of copper chloride and fuller's earth. It applies to both kerosene and gasoline. The oil is heated and brought into contact with the slurry while being agitated in a stream of air that oxidizes the mercaptans to disulfides. The slurry is then allowed to settle and is separated for reuse. A heater raises the temperature to a point that keeps the water formed in the reaction dissolved in the oil so that the catalyst remains properly hydrated. After sweetening, the oil is water-washed to remove any traces of the catalyst and is later dried by passing through a salt filter.

4.15   MERCAPTAN EXTRACTION

Simple sweetening is adequate for many purposes, but other methods must be used if the total sulfur content of the fuel is to be reduced. When solutizers, such as potassium isobutyrate and sodium cresylate, are added to caustic soda, the solubility of the higher mercaptans is increased and they can be extracted from the oil. To remove traces of hydrogen sulfide and alkyl phenols, the oil is first pretreated with caustic soda in a packed column or other mixing device. The mixture is allowed to settle and the product water washed before storage.

4.16   CLAY TREATMENT

Some natural clays, activated by roasting or treatment with steam or acids, have been used for many years to remove traces of impurities. The phenomenon is similar to that described under the adsorption process: the clay retains the longer chain molecules within its highly porous structure.

Clay treatment removes gum and gum-forming materials from thermally cracked gasoline in the vapor phase. A more economical procedure, however, is to add small quantities of synthetic antioxidants to the gasoline. These prevent or greatly retard gum formation. Clay treatment of lubricating oils is widely practiced to remove resins and other color bodies remaining after solvent extraction. The treatment may be by contact—that is, clay added directly to the oil, with the mixture heated and the clay filtered off—or by percolation, in which the heated oil is passed through a large bed of active clay adsorbent. The spent clay is often discarded, although it can be regenerated by roasting. However, the problem of dealing with spent clay, now designated as hazardous waste in many places, has led many refiners to replace clay treatment facilities with a mild hydrogenation process.

4.17   HYDROGEN TREATMENT

Hydrogen processes, commonly known as hydrotreating, are the most common processes for removing sulfur and nitrogen impurities. The oil is combined with high-purity hydrogen, vaporized, and then passed over a catalyst such as tungsten, nickel, or a mixture of cobalt and molybdenum oxides supported on an alumina base. Operating temperatures are usually between 260 and 425 °C (500 and 800 °F) at pressures of 14 to 70 bars (1.4 to 7 MPa), or 200 to 1,000 psi. Operating conditions are set to facilitate the desired level of sulfur removal without promoting any change to the other properties of the oil.

The sulfur in the oil is converted to hydrogen sulfide and the nitrogen to ammonia. The hydrogen sulfide is removed from the circulating hydrogen stream by absorption in a solution such as diethanolamine. The solution can then be heated to remove the sulfide and reused. The hydrogen sulfide recovered is useful for manufacturing elemental sulfur of high purity. The ammonia is recovered and either converted to elemental nitrogen and hydrogen, burned in the refinery fuel-gas system, or processed into agricultural fertilizers.

4.18   MOLECULAR SIEVES

Molecular sieves are also used to purify petroleum products since they have a strong affinity for polar compounds such as water, carbon dioxide, hydrogen sulfide, and mercaptans. Sieves are prepared by dehydration of an aluminosilicate such as zeolite. The petroleum product is passed through a bed of zeolite for a predetermined period depending on the impurity to be removed. The adsorbed contaminants may later be expelled from the sieve by purging with a gas stream at temperatures between 200 and 315 °C (400 and 600 °F). The frequent cycling of the molecular sieve from adsorb to desorb operations is usually fully automated.

CHAPTER 5

PETROLEUM PRODUCTS AND THEIR USES

5.1   GASES

Gaseous refinery products include hydrogen, fuel gas, ethane, propane, and butane. Most of the hydrogen is consumed in refinery desulfurization facilities, which remove hydrogen sulfide from the gas stream and then separate that compound into elemental hydrogen and sulfur; small quantities of the hydrogen may be delivered to the refinery fuel system. Refinery fuel gas varies in composition but usually contains a significant amount of methane; it has a heating value similar to natural gas and is consumed in plant operations. Periodic variability in heating value makes it unsuitable for delivery to consumer gas systems. Ethane may be recovered from the refinery fuel system for use as a petrochemical feedstock. Propane and butane are sold as liquefied petroleum gas (LPG), which is a convenient portable fuel for domestic heating and cooking or light industrial use.

5.2   GASOLINE

Motor gasoline, or petrol, must meet three primary requirements. It must provide an even combustion pattern, start easily in cold weather, and meet prevailing environmental requirements.

5.3   OCTANE RATING

To meet the first requirement, gasoline must burn smoothly in the engine without premature detonation, or knocking. Severe knocking can dissipate power output and even cause damage to the engine. When gasoline engines became more powerful in the 1920s, it was discovered that some fuels knocked more readily than others. Experimental studies led to the determination that, of the standard fuels available at the time, the most extreme knock was produced by a fuel composed of pure normal heptane, while the least knock was produced by pure isooctane. This discovery led to the development of the octane scale for defining gasoline quality. Thus, when a motor gasoline gives the same performance in a standard knock engine as a mixture of 90 percent isooctane and 10 percent normal heptane, it is given an octane rating of 90.

There are two methods for carrying out the knock engine test. Research octane is measured under mild conditions of temperature and engine speed (49 °C [120 °F] and 600 revolutions per minute, or RPM), while motor octane is measured under more severe conditions (149 °C [300 °F] and 900 RPM). For many years the research octane number was found to be the more accurate measure of engine performance and was usually quoted alone. Since the advent of unleaded fuels in the mid-1970s, however, motor octane measurements have frequently been found to limit actual engine performance. As a result a new measurement, road octane number, which is a simple average of the research and motor values, is most frequently used to define fuel quality for the consumer. Automotive gasoline generally ranges from research octane number 87 to 100, while gasoline for piston-engine aircraft ranges from research octane number 115 to 130.

Each naphtha component that is blended into gasoline is tested separately for its octane rating. Reformate, alkylate, polymer, and cracked naphtha, as well as butane, all rank high (90 or higher) on this scale, while straight-run naphtha may rank at 70 or less. In the 1920s it was discovered that the addition of tetraethyl lead would substantially enhance the octane rating of various naphthas. Each naphtha component was found to have a unique response to lead additives, some combinations being found to be synergistic and others antagonistic. This gave rise to very sophisticated techniques for designing the optimal blends of available components into desired grades of gasoline.

The advent of led, or ethyl, gasoline led to the manufacture of high-octane fuels and became universally employed throughout the world after World War II. However, beginning in 1975, environmental legislation began to restrict the use of lead additives in automotive gasoline. It is now banned in the United States, the European Union, and many countries around the world. The required use of lead-free gasoline has placed a premium on the construction of new catalytic reformers and alkylation units for increasing yields of high-octane gasoline ingredients and on the exclusion of low-octane naphthas from the gasoline blend.

5.4   HIGH-VOLATILE AND LOW-VOLATILE COMPONENTS

The second major criterion for gasoline—that the fuel be sufficiently volatile to enable the car engine to start quickly in cold weather—is accomplished by the addition of butane, a very low-boiling paraffin, to the gasoline blend. Fortunately, butane is also a high-octane component with little alternate economic use, so its application has historically been maximized in gasoline. Another requirement, that a quality gasoline has a high energy content, has traditionally been satisfied by including higher-boiling components in the blend. However, both of these practices are now called into question on environmental grounds. The same high volatility that provides good starting characteristics in cold weather can lead to high evaporative losses of gasoline during refueling operations, and the inclusion of high-boiling components to increase the energy content of the gasoline can also increase the emission of unburned hydrocarbons from engines on start-up. As a result, since the 1990 amendments to the U.S. Clean Air Act, much of the gasoline consumed in urban areas of the United States has been reformulated to meet stringent new environmental standards. At first, these changes required that gasoline contain certain percentages of oxygen to aid in fuel combustion and reduce the emission of carbon monoxide and nitrogen oxides. Refiners met this obligation by including some oxygenated compounds such as ethyl alcohol or methyl tertiary butyl ether (MTBE) in their blends. However, MTBE was soon judged to be a hazardous pollutant of groundwater in some cases where reformulated gasoline leaked from transmission pipelines or underground storage tanks, and it was banned in several parts of the country. In 2005 the requirements for specific oxygen levels were removed from gasoline regulations, and MTBE ceased to be used in reformulated gasoline. Many blends in the United States contain significant amounts of ethyl alcohol to meet emissions requirements, and MTBE is still added to gasoline in other parts of the world.

5.5   GASOLINE BLENDING

One of the most critical economic issues for a petroleum refiner is selecting the optimal combination of components to produce final gasoline products. Gasoline blending is much more complicated than a simple mixing of components. First, a typical refinery may have as many as 8 to 15 different hydrocarbon streams to consider as blend stocks. These may range from butane, the most volatile component, to heavy naphtha and include several gasoline naphthas from crude distillation, catalytic cracking, and thermal processing units in addition to alkylate polymer and reformate. Modern gasoline may be blended to meet simultaneously 10 to 15 different quality specifications, such as vapor pressure; initial, intermediate, and final boiling points; sulfur content; color; stability; aromatics content; olefin content; octane measurements for several different portions of the blend; and other local governmental or market restrictions. Since each of the individual components contributes uniquely to each of these quality areas and each bears a different cost of manufacture, the proper allocation of each component into its optimal disposition is of major economic importance. To address this problem, most refiners employ linear programming, a mathematical technique that permits the rapid selection of an optimal solution from a multiplicity of feasible alternative solutions. Each component is characterized by its specific properties and cost of manufacture, and each gasoline grade requirement is similarly defined by quality requirements and relative market value. The linear programming solution specifies the unique disposition of each component to achieve maximum operating profit. The next step is to measure carefully the rate of addition of each component to the blend and collect it in storage tanks for final inspection before delivering it for sale. Still, the problem is not fully resolved until the product is delivered to customers' tanks. Frequently, last-minute changes in shipping schedules or production qualities require the reblending of finished gasoline or the substitution of a high-quality (and therefore costlier) grade for one of more immediate demand even though it may generate less income for the refinery.

5.6   KEROSENE

Though its use as an illuminant has greatly diminished, kerosene is still used extensively throughout the world in cooking and space heating and is the primary fuel for modern jet engines. When burned as a domestic fuel, kerosene must produce a flame free of smoke and odor. Standard laboratory procedures test these properties by burning the oil in special lamps. All kerosene fuels must satisfy minimum flash-point specifications (49 °C, or 120 °F) to limit fire hazards in storage and handling.

Jet fuels must burn cleanly and remain fluid and free from wax particles at the low temperatures experienced in high-altitude flight. The conventional freeze-point specification for commercial jet fuel is −50 °C (−58 °F). The fuel must also be free of any suspended water particles that might cause blockage of the fuel system with ice particles. Special-purpose military jet fuels have even more stringent specifications.

5.7 DIESEL OILS

The principal end use of gas oil is as diesel fuel for powering automobiles, trucks, buses, and railway engines. In a diesel engine, combustion is induced by the heat of compression of the air in the cylinder under compression. Detonation, which leads to harmful knocking in a gasoline engine, is a necessity for the diesel engine. A good diesel fuel starts to burn at several locations within the cylinder after the fuel is injected. Once the flame has initiated, any more fuel entering the cylinder ignites at once.

Straight-chain hydrocarbons make the best diesel fuels. To have a standard reference scale, the oil is matched against blends of cetane (normal hexadecane) and alpha methylnaphthalene, the latter of which gives very poor engine performance. High-quality diesel fuels have cetane ratings of about 50, giving the same combustion characteristics as a 50-50 mixture of the standard fuels. The large, slower engines in ships and stationary power plants can tolerate even heavier diesel oils. The more viscous marine diesel oils are heated to permit easy pumping and to give the correct viscosity at the fuel injectors for good combustion.

Until the early 1990s, standards for diesel fuel quality were not particularly stringent. A minimum cetane number was critical for transportation uses, but sulfur levels of 5,000 parts per million (ppm) were common in most markets. With the advent of more stringent exhaust emission controls, however, diesel fuel qualities came under increased scrutiny. In the European Union and the United States, diesel fuel is now generally restricted to maximum sulfur levels of 10 to 15 ppm, and regulations have restricted aromatic content as well. The limitation of aromatic compounds requires a much more demanding scheme of processing individual gas oil components than was necessary for earlier highway diesel fuels.

5.8   FUEL OILS

Furnace oil consists largely of residues from crude oil refining. These are blended with other suitable gas oil fractions to achieve the viscosity required for convenient handling. As a residue product, fuel oil is the only refined product of significant quantity that commands a market price lower than the cost of crude oil.

Because the sulfur contained in the crude oil is concentrated in the residue material, fuel oil sulfur levels are naturally high. The sulfur level is not critical to the combustion process as long as the flue gases do not impinge on cool surfaces (which could lead to corrosion by the condensation of acidic sulfur trioxide). However, to reduce air pollution, most industrialized countries now restrict the sulfur content of fuel oils. Such regulation has led to the construction of residual desulfurization units or cokers in refineries that produce these fuels.

Residual fuels may contain large quantities of heavy metals such as nickel and vanadium; these produce ash upon burning and can foul burner systems. Such contaminants are not easily removed and usually lead to lower market prices for fuel oils with high metal contents.

5.9   LUBRICATING OILS

At one time the suitability of petroleum fractions for use as lubricants depended entirely on the crude oils from which they were derived. Those from Pennsylvania crude, which were largely paraffinic, were recognized as having superior properties. But, with the advent of solvent extraction and hydrocracking, the choice of raw materials has been considerably extended.

Viscosity is the basic property by which lubricating oils are classified. The requirements vary from a very thin oil needed for the high-speed spindles of textile machinery to the viscous, tacky materials applied to open gears or wire ropes. Between these extremes is a wide range of products with special characteristics. Automotive oils represent the largest product segment in the market. In the United States, specifications for these products are defined by the Society of Automotive Engineers (SAE), which issues viscosity ratings with numbers that range from 5 to 50. In the United Kingdom, standards are set by the Institute of Petroleum, which conducts tests that are virtually identical to those of the SAE.

When ordinary mineral oils having satisfactory lubricity at low temperatures are used over an extended temperature range, excessive thinning occurs, and the lubricating properties are found to be inadequate at higher temperatures. To correct this, multigrade oils have been developed using long-chain polymers. Thus, an oil designated SAE 10W40 has the viscosity of an SAE 10W oil at −18 °C (0 °F) and of an SAE 40 oil at 99 °C (210 °F). Such an oil performs well under cold starting conditions in winter (hence the W designation) yet will lubricate under high-temperature running conditions in the summer as well. Other additives that improve the performance of lubricating oils are antioxidants and detergents, which maintain engine cleanliness and keep fine carbon particles suspended in the circulating oil.

5.9   GEAR OILS AND GREASES

In gear lubrication the oil separates metal surfaces, reducing friction and wear. Extreme pressures develop in some gears, and special additives must be employed to prevent the seizing of the metal surfaces. These oils contain sulfur compounds that form a resistant film on the surfaces, preventing actual metal-to-metal contact.

Greases are lubricating oils to which thickening agents are added. Soaps of aluminum, calcium, lithium, and sodium are commonly used, while nonsoap thickeners such as carbon, silica, and polyethylene also are employed for special purposes.

5.10   OTHER PETROLEUM PRODUCTS

Highly purified naphthas are used for solvents in paints, cosmetics, commercial dry cleaning, and industrial product manufacture. Petroleum waxes are employed in paper manufacture and foodstuffs.

Asphaltic bitumen is widely used for the construction of roads and airfields. Specialized applications of bitumen also include the manufacture of roofing felts, waterproof papers, pipeline coatings, and electrical insulation. Carbon black is manufactured by decomposing liquid hydrocarbon fractions. It is compounded with rubber in tire manufacture and is a constituent of printing inks and lacquers.

5.11   PETROCHEMICALS

By definition, petrochemicals are simply chemicals that happen to be derived from a starting material obtained from petroleum. They are, in almost every case, virtually identical to the same chemical produced from other sources, such as coal, coke, or fermentation processes.

5.12   OLEFINS

The thermal cracking processes developed for refinery processing in the 1920s were focused primarily on increasing the quantity and quality of gasoline components. As a by-product of this process, gases were produced that included a significant proportion of lower-molecular-weight olefins, particularly ethylene, propylene, and butylene. Catalytic cracking is also a valuable source of propylene and butylene, but it does not account for a very significant yield of ethylene, the most important of the petrochemical building blocks. Ethylene is polymerized to produce polyethylene or, in combination with propylene, to produce copolymers that are used extensively in food-packaging wraps, plastic household goods, or building materials.

Ethylene manufacture via the steam cracking process is in widespread practice throughout the world. The operating facilities are similar to gas oil cracking units, operating at temperatures of 840 °C (1,550 °F) and at low pressures of 165 kilopascals (24 pounds per square inch). Steam is added to the vaporized feed to achieve a 50-50 mixture, and furnace residence times are only 0.2 to 0.5 seconds. In the United States and the Middle East, ethane extracted from natural gas is the predominant feedstock for ethylene cracking units. Propylene and butylene are largely derived from catalytic cracking units in the United States. In Europe and Japan, catalytic cracking is less common, and natural gas supplies are not as plentiful. As a result, both the Europeans and Japanese generally crack a naphtha or light gas oil fraction to produce a full range of olefin products.

5.13   AROMATICS

The aromatic compounds, produced in the catalytic reforming of naphtha, are major sources of petrochemical products. In the traditional chemical industry, aromatics such as benzene, toluene, and xylenes were made from coal during carbonization in the production of coke and town gas. Today a much larger volume of these chemicals are made as refinery by-products.

A further source of supply is the aromatic-rich liquid fraction produced in the cracking of naphtha or light gas oils during the manufacture of ethylene and other olefins.

5.14   POLYMERS

A highly significant proportion of these basic petrochemicals is converted into plastics, synthetic rubbers, and synthetic fibers. Together these materials are known as polymers, because their molecules are high-molecular-weight compounds made up of repeated structural units that have combined chemically. The major products are polyethylene, polyvinyl chloride, and polystyrene, all derived from ethylene, and polypropylene, derived from monomer propylene. Major raw materials for synthetic rubbers include butadiene, ethylene, benzene, and propylene. Among synthetic fibres the polyesters, which are a combination of ethylene glycol and terephthalic acid (made from xylenes), are the most widely used. They account for about one-half of all synthetic fibers. The second major synthetic fiber is nylon, its most important raw material being benzene. Acrylic fibers, in which the major raw material is the propylene derivative acrylonitrile, make up most of the remainder of the synthetic fibers.

5.15   INORGANIC CHEMICALS

Two prominent inorganic chemicals, ammonia and sulfur, are also derived in large part from petroleum. Ammonia production requires hydrogen from a hydrocarbon source. Traditionally, hydrogen was produced from a coke and steam reaction, but today most ammonia is synthesized from liquid petroleum fractions, natural gas, or refinery gases. The sulfur removed from oil products in purification processes is ultimately recoverable as elemental sulfur or sulfuric acid. It has become an important source of sulfur for the manufacture of fertilizer.

CHAPTER 6

OIL REFINERY

An oil refinery or Petroleum refinery is an industrial process plant where petroleum (crude oil) is transformed and refined into products such as gasoline (petrol), diesel fuel, asphalt base, fuel oils, heating oil, kerosene, liquefied petroleum gas and petroleum naphtha. Petrochemical feedstock like ethylene and propylene can also be produced directly by cracking crude oil without the need of using refined products of crude oil such as naphtha. The crude oil feedstock has typically been processed by an oil production plant. There is usually an oil depot at or near an oil refinery for the storage of incoming crude oil feedstock as well as bulk liquid products. In 2020, the total capacity of global refineries for crude oil was about 101.2 million barrels per day.

Fig 6.1: Refinery plant and facilities

6.2 PROCESSING CONFIGURATIONS

Each petroleum refinery is uniquely configured to process a specific raw material into a desired slate of products. To determine which configuration is most economical, engineers and planners survey the local market for petroleum products and assess the available raw materials. Since about half the product of fractional distillation is residual fuel oil, the local market for it is of utmost interest. In parts of Africa, South America, and Southeast Asia, heavy fuel oil is easily marketed, so that refineries of simple configuration may be sufficient to meet demand. However, in the United States, Canada, and Europe, large quantities of gasoline are in demand, and the market for fuel oil is constrained by environmental regulations and the availability of natural gas. In these places, more complex refineries are necessary.

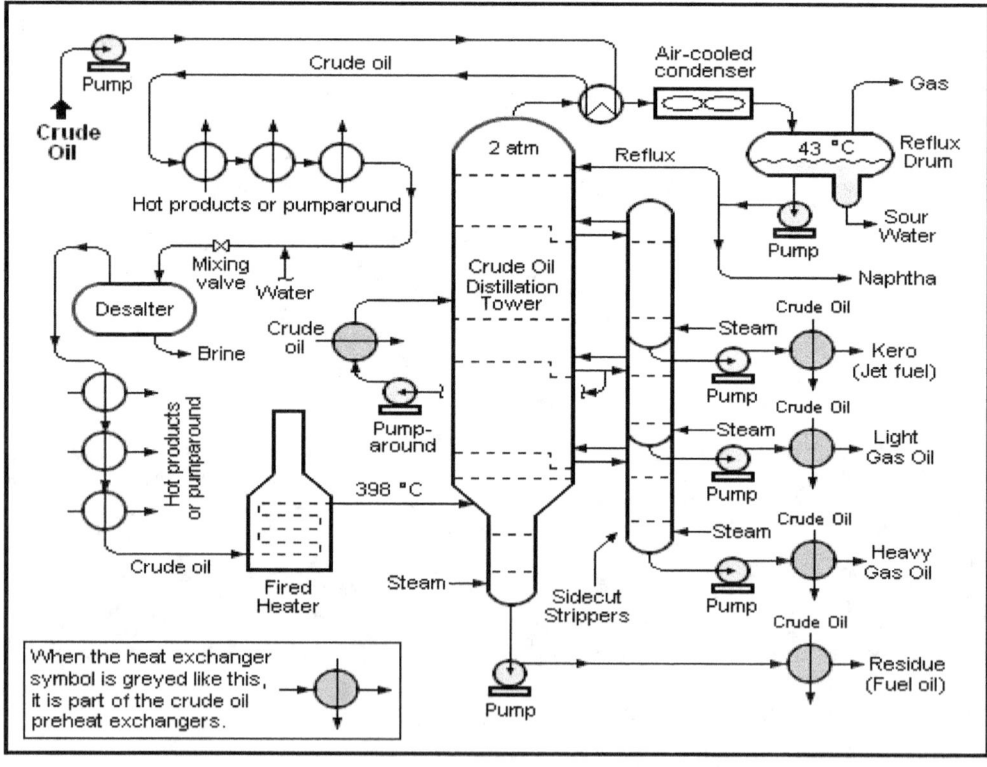

6.3 PETROLEUM REFINING PROCESSES

6.3.1 Topping and hydroskimming refineries

The simplest refinery configuration called a topping refinery, is designed to prepare feedstocks for petrochemical manufacture or production of industrial fuels in remote oil-production areas. It consists of tankage, a distillation unit, recovery facilities for gases and light hydrocarbons, and the necessary utility systems (steam, power, and water-treatment plants).

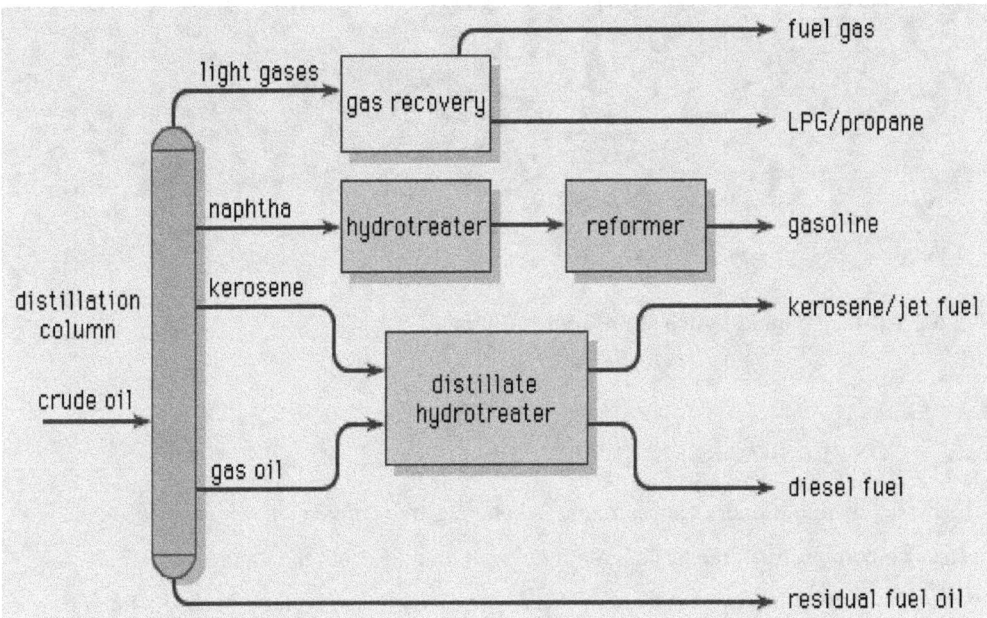

Fig 7.3.1 (a): Unit operation in a hydroskimming refinery

Fig 6.3.1 (b): A typical hydro skimming refinery

6.4   CONVERSION REFINERY

The most versatile refinery configuration is known as the conversion refinery. A conversion refinery incorporates all the basic building blocks found in both the topping and hydroskimming refineries, but it also features gas oil conversion plants such as catalytic cracking and hydrocracking units, olefin conversion plants such as alkylation or polymerization units, and, frequently, coking units for sharply reducing or eliminating the production of residual fuels. Modern conversion refineries may produce two-thirds of their output as gasoline, with the balance distributed between high-quality jet fuel, liquefied petroleum gas (LPG), diesel fuel, and a small quantity of petroleum coke. Many such refineries also incorporate solvent extraction processes for manufacturing lubricants and petrochemical units with which to recover high-purity propylene, benzene, toluene, and xylenes for further processing into polymers.

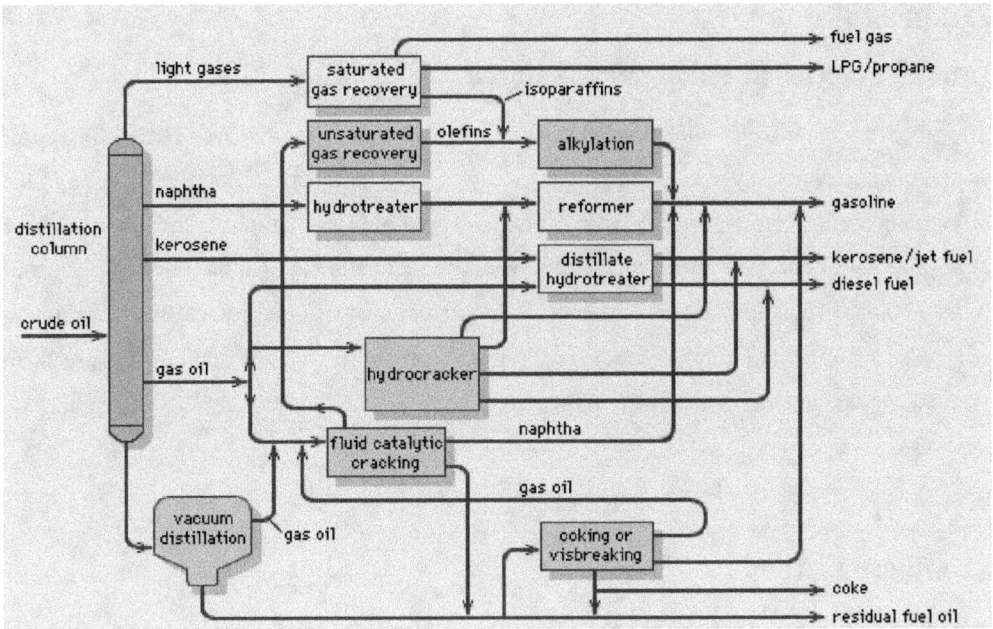

Fig 6.4: Unit operation in conversion refinery

6.5   OFF-SITES

The individual processing units described above are part of the process-unit side of a refinery complex. They are usually considered the most important features, but the functioning of the off-site facilities is often as critical as the process units themselves. Off-sites consist of tankage, flare systems, utilities, and environmental treatment units.

CHAPTER 7

OIL REFINING AND THE ENVIRONMENT

The impact of man's activities can upset the ecosystem in several ways. The discharge of pollutants into water streams can reduce their capacity for self-purification thus, results in the elimination of species of organisms and plants. The emission of toxic compounds into the atmosphere will not only affect vegetation, birds, and animals but also the physical health of human beings. Thus, it is essential to control the activities of man to protect the environment. Air, water, and land are vital to life on earth. Man must protect these resources and use them sustainably as our survival depends on them.

7.1 AIR POLLUTION

Any process that adds or subtracts from the usual constituents of air may alter its physical or chemical properties sufficiently to be detected by occupants of the medium. Therefore, pollutants include any natural or artificial composition of matter capable of being airborne – solid particulates, liquid droplets, gases, or various admixtures of these forms.

Pollutants can be classified into two:

Primary pollutants: those emitted directly from identifiable sources. They include particulate matter, sulfur compounds, radioactive compounds, etc.

Secondary pollutants: those formed in the air by interaction between various primary pollutants or by reaction with normal atmosphere constituents. Examples are the formation of sulphuric acid mist, smoke, etc.

Effects of air pollution

Pollutants in the atmosphere have several effects on the environment. They include:

➢ Reduction in visibility
➢ Damage to the materials

- Damage to vegetation
- Physiological effects on men and animals
- Psychological effects
- The Oil Pollution Act

The Oil Pollution Act of 1990 deals with pollution of waterways by crude oil. The Act specifically deals with petroleum vessels and onshore and offshore facilities and imposes strict liability for oil spills on their owners and operators.

7.2   EMISSIONS AND ENVIRONMENTAL EFFECTS

Globally, the production, distribution, and consumption of hydrocarbons as fuel or feedstock are the largest source of emissions into the environment. About 81% of the total annual world energy supply of 11,000 million tons of oil equivalent is based on fossil fuels, which emit about 26,000 million tons of carbon dioxide and other gases, such as methane into the atmosphere.

The most serious effect of these emissions is global climate change. The Intergovernmental Panel on Climate Change (regularly called the UN Climate Panel) predicts that these emissions will cause the global temperature to rise from between 1.4 to 6.4 °C by the end of the 21st century, depending on models and global scenarios.

7.3   THE HAZARDOUS MATERIALS TRANSPORTATION ACT

The Hazardous Materials Transportation Act authorizes the establishment and enforcement of hazardous material regulations for all modes of transportation by highway, water, and rail. The purpose of the Act is to ensure the safe transportation of hazardous materials. The Act prevents any person from offering or accepting a hazardous material for transportation anywhere within this nation if that material is not properly classified, described, packaged, marked, labeled, and authorized for shipment under the regulatory requirements.

Under Department of Transportation regulations, a hazardous material is defined as any substance or material, including a hazardous substance and hazardous waste, which is capable of posing an unreasonable risk to health, safety, and property during transportation. The Act also imposes restrictions on the packaging, handling, and shipping of hazardous materials. For shipping and receiving hazardous chemicals, hazardous wastes, and radioactive materials, the appropriate documentation, markings, labels, and safety precautions are required.

7.4   ENVIRONMENTAL PROTECTION

Environmental problems arise in dealing with crude oil and its products. The petroleum industry maintains exploration, transport, and refining installations for crude oil. Since the geographical location of crude oil fields does not happen together with that of the consuming and refining regions, enormous transportation distances exist. This means that the industry needs to take precautions in all areas about air emissions, water and soil pollution, and, to a lesser degree, noise. The cost of technical measures for environmental conservation amounts to more than 25% of the total processing costs in the refineries. The environmental problems in crude oil refining, storage, and loading and the application of the products are discussed below.

7.5   PETROLEUM REFINING WASTE

Petroleum refining produces chemical waste. If this chemical waste is not processed promptly, it can become a pollutant. A pollutant is a substance present in a particular location (ecosystem) when it is not indigenous to the location or is present in a greater-than-natural concentration. The substance is often the product of human activity. The pollutant has a detrimental effect on the environment, in part or total.

Pollutants are subdivided into two classes: primary and secondary.

Source → Primary pollutant → Secondary pollutant

1. A primary pollutant: is a pollutant that is emitted directly from the source. In terms of atmospheric pollutants, examples are carbon oxides, sulfur dioxide, and nitrogen oxides from fuel combustion operations:

- $2[C]$petroleum $+ O_2 \rightarrow 2CO$
- $[C]$petroleum $+ O_2 \rightarrow CO_2$
- $2[N]$petroleum $+ O_2 \rightarrow 2NO$
- $[N]$petroleum $+ O_2 \rightarrow NO_2$
- $[S]$petroleum $+ O_2 \rightarrow SO_2$
- $2SO_2 + O_2 \rightarrow 2SO_3$ or hydrogen sulfide and ammonia from processing sulfur-containing raw materials:
- $[S]$petroleum $+ H_2 \rightarrow H_2S +$ hydrocarbons
- $2[N]$petroleum $+ 3H_2 \rightarrow 2NH_3 +$ hydrocarbons

The question of classifying nitrogen dioxide and sulfur trioxide as primary pollutants often arises, as does the origin of the nitrogen. In the former case, these higher oxides can be formed in the upper levels of the combustion reactors.

A secondary pollutant: is a pollutant that is produced by the interaction of a primary pollutant with another chemical. A secondary pollutant may also be produced by dissociation of a primary pollutant, or other effects within a particular ecosystem. Again, using the atmosphere as an example, the formation of the constituents of acid rain is an example of the formation of secondary pollutants:

- $SO_2 + H_2O \rightarrow H_2SO_3$ (sulphurous acid)
- $SO_3 + H_2O \rightarrow H_2SO_4$ (sulphuric acid)
- $NO + H_2O \rightarrow HNO_2$ (nitrous acid)
- $3NO_2 + 2H_2O \rightarrow HNO_3$ (nitric acid)

In many cases, these secondary pollutants can have significant environmental effects, such as the formation of acid rain and smog.

Any pollutant, either primary or secondary can have a serious effect on the various ecological cycles. Therefore, understanding how a chemical pollutant can enter these ecosystems and influence the future behavior of the ecosystem, is extremely important.

In addition, hazardous waste is any gaseous, liquid, or solid waste material that, if improperly managed or disposed of, may pose substantial hazards to human health and the environment.

In many cases, the term chemical waste is often used interchangeably with the term hazardous waste. However, not all chemical wastes are hazardous and caution in the correct use of the terms must be exercised lest unqualified hysteria take control.

7.6   ENVIRONMENTAL REGULATIONS

Environmental issues permeate everyday life. These issues range from the effects on the lives of workers in various occupations where hazards can result from exposure to chemical agents to the influence of these agents on the lives of the population at large In this section, reference is made to the various environmental laws.

An environmental regulation is a legal mechanism that determines how a statute's broad policy directives are to be carried out. An environmental policy is a requirement that specifies operating procedures that must be followed. An environmental guidance is a document developed by a governmental agency that outlines a position on a topic or gives instructions on how a procedure must be carried out. It explains how to do something and provides governmental interpretations of a governmental act or policy.

7.7   THE WATER POLLUTION CONTROL ACT (THE CLEAN WATER ACT)

The objective of the Water Pollution Control Act (Clean Water Act) is to restore and maintain the chemical, physical, and biological integrity of water systems. The Water Quality Act is aimed at improving water quality in areas where there were insufficiencies in compliance with the discharge standards.

Section 311 of the Clean Water Act includes elaborate provisions for regulating intentional or accidental discharges of petroleum and hazardous substances. Included are response actions required for oil spills and the release or discharge of toxic and hazardous substances. As an example, the person in charge of a vessel or an onshore or offshore facility from which any designated hazardous substance is discharged, in quantities equal to or exceeding its reportable quantity, must notify the appropriate federal agency as soon as such knowledge is obtained.

The Exxon Valdez is a well-known case.

7.8   THE SAFE DRINKING WATER ACT

The Safe Drinking Water Act, first enacted in 1974, was amended several times in the 1970s and 1980s to set national drinking water standards. The Act calls for regulations that

Apply to public water systems,

Specify contaminants that may have any adverse effect on the health of persons, and

Specify contaminant levels. In addition, the difference between primary and secondary drinking water regulations is defined, and a variety of analytical procedures is specified.

Statutory provisions are included to cover underground injection control systems. The Act also requires maximum levels at which a contaminant must have no known or anticipated adverse effects on human health, thereby providing an adequate margin of safety.

The Superfund Amendments and Reauthorization Act (SARA) set the same standards for groundwater as for drinking water in terms of necessary clean-up and remediation of an inactive site that might be a former petroleum refinery. Under the Act, all underground injection activities must comply with the drinking water standards as well as meet specific permit conditions that are in unison with the provisions of the Clean Water Act. However, under the

Resource Conservation and Recovery Act, class IV injection wells are no longer permitted and several restrictions on underground injection wells may be used for storage and disposal of hazardous wastes.

7.9   THE RESOURCE CONSERVATION AND RECOVERY ACT

Since its initial enactment in 1976, the Resource Conservation and Recovery Act (RCRA) continues to promote safer solid and hazardous waste management programs. Besides the regulatory requirements for waste management, the Act specifies the mandatory obligations of generators, transporters, and disposers of waste as well as those of owners and/or operators of waste treatment, storage, or disposal facilities. The Act also defines solid waste as garbage, refuse, sludge from a treatment plant, from a water supply treatment plant, or air pollution control facility, and other discarded material, including solid, liquid, semisolid, or contained gaseous material resulting from industrial, commercial, mining, and agricultural operations and from community activities.

The Act also states that solid waste does not include solid, or dissolved, materials in domestic sewage, or solid or dissolved materials in irrigation return flows or industrial discharges. A solid waste becomes a hazardous waste if it exhibits any one of four specific characteristics: (1) ignitability, (2) reactivity, (3) corrosivity, or (4) toxicity. Certain types of solid wastes (e.g., household waste) are not considered to be hazardous irrespective of their characteristics.

Hazardous waste generated in a product or raw-material storage tank, transport vehicles, or manufacturing processes and samples collected for monitoring and testing purposes are exempt from the regulations. Hazardous waste management is based on a beginning-to-end concept so that all hazardous wastes can be traced and fully accounted for. All generators and transporters of hazardous wastes as well as owners and operators of related facilities in the United States must file a notification with the Environmental Protection Agency. The notification must state the location of the facility and a general description of the activities as well as the identified and listed hazardous wastes being handled. Thus, all regulated hazardous waste facilities must exist and/or operate under valid, activity-specific permits.

7.10   THE TOXIC SUBSTANCES CONTROL ACT

The Toxic Substances Control Act was first enacted in 1976 and was designed to provide controls for those chemicals that may threaten human health or the environment. Particularly hazardous are the cyclic nitrogen species that may be produced when petroleum is processed and that often occur in residua and cracked residua. The objective of the Act is to provide the necessary control before a chemical is allowed to be mass-produced and enter the environment.

The Act specifies a premanufactured notification requirement by which any manufacturer must notify the Environmental Protection Agency at least 90 days before the production of a new chemical substance. Notification is also required even if there is a new use for the chemical that can increase the risk to the environment. No notification is required for chemicals that are manufactured in small quantities solely for scientific research and experimentation. A new chemical substance is defined as a chemical that is not listed in the Environmental Protection Agency

7.11   THE COMPREHENSIVE ENVIRONMENTAL RESPONSE, COMPENSATION, AND LIABILITY ACT

The Comprehensive Environmental Response, Compensation, and Liability Act (CERCLA) which is generally known as Superfund, was first signed into law in 1980. The central purpose of this Act is to provide a response mechanism for clean-up of any hazardous substance released, such as an accidental spill, or of a threatened release of a chemical. While RCRA deals basically with the management of wastes that are generated, treated, stored, or disposed of, CERCLA responds to the environmental release of various pollutants or contaminants into the air, water, or land.

1. Manufacturing Emissions: Though closed, gas-tight systems are used in refinery units, emissions into air and water cannot be completely avoided even with careful handling during refining and storage of the crude oil and its products. Hydrocarbons are discharged into the air because of their high vapor pressure and they appear in refinery wastewater effluents because of their water solubility, which is, however, small. The carcinogenic aromatic hydrocarbons are particularly dangerous.

Further attention must be paid to the sulfur and nitrogen compounds originating from the heteroatomic compounds in the crude oil, both because of their smell and toxicity and because of the air pollution, which arises in the form of SO_2 and NO_x emissions during firing in process plants.

1. Hydrocarbons (HC) in Air: Hydrocarbon emissions can arise in production plants during normal operation:

(1) From leaking flanges in the pipework system;

(2) At the seals of valves, pumps, and compressors; and

(3) In the course of sampling.

In the case of an accident, gases are led in closed systems to flares, collected as far as possible in gas recovery systems, compressed, and returned to the process. The remainder of the gases is burnt in elevated or ground flares at efficiencies, which can exceed 99 %. Liquid products are collected in closed slop systems that are equipped with pressure reservoirs and tanks and later returned to the production circuit.

In new plants, measures to reduce emissions are taken during their construction, whereas continuous retrofitting is necessary in existing plants. Examples are flangeless piping, low-emission stuffing boxes, and seals, such as duplicated slide ring packings. For intensely odorous, poisonous, and carcinogenic substances, more far-reaching measures are necessary (canned motor pumps, special extraction devices, etc.).

Most hydrocarbon emissions in processing occur in the storage areas, i.e., tank farms for crude, feedstocks, intermediate, and final products. Pressure–vacuator relief valves and floating cover tanks are generally used to reduce emissions. More developments that are recent are emission-free tank farms, where several fixed roof tanks fitted with internal floating roofs breathe into a closed system at one common gasometer, which normally absorbs all changes in the tank level. In the rare event of unusually large changes in the system (large amounts flowing into or out of storage, solar irradiation, heavy rain), the surplus quantity is burnt harmlessly in an associated flare or, in the case of a pressure drop, the system is topped up with inert gas.

Considerable amounts of hydrocarbons are also emitted in the loading facilities, especially in the loading of gasoline. Here low-emission or emission-free loading for transport by ship, rail, and road has become largely accepted. Various methods used are:

i.   Vapor recovery, where in a closed system the displaced gasoline vapors are either returned to the product tank or collected in a gasometer and used for the firing of process plants.

ii.   Regenerative adsorption of the vapors on suitable adsorbents

iii.   Recovery of the products in liquid form after cooling or washing out the vapors.

Care must be taken to prevent the formation of explosive gasoline–air mixtures that can occur in the road tanker to be loaded and in the adjacent piping and equipment.

This can be achieved by:

a.   Keeping the concentrations outside the explosive range,
b.   Short transportation paths exclusion of ignition sources, and
c.   Extremely strict control of the oxygen contents.

Dispensing of gasoline from the road tanker to the service station is also increasingly carried out with vapor recovery between the road tanker and the installed storage tank of the service station.

7.12   HYDROCARBONS IN WASTEWATER

Hydrocarbon-containing wastewater is unavoidably obtained at various points in the refinery because of the water content of the crude oil itself and because steam is employed in various processing steps. The total amount of hydrocarbon-containing wastewater in a normal refinery is of the order of 60 – 100m3/h. This wastewater must be removed from the process units after separation of the oil phase and led via a closed system to wastewater purification.

Rainwater from exposed plant areas and tank yards, and possibly contaminated cooling water from leaks or accidents must be treated in the same way. Considerable buffer volumes must be made available for the latter amounts of water, which are formed discontinuously, and sometimes in large quantities.

Treatment in the wastewater purification system is carried out stepwise by:

- Mechanical separation (sieves, filters, oil–water separators)
- Physicochemical purification (stripping, flocculation, flotation)
- Biological treatment

Biological wastewater treatment of hydrocarbons in refinery wastewater is normally problem-free. However, the incoming streams and corresponding buffer volume must be continuously monitored to detect pollution by sulfur or nitrogen compounds and by oxygen-containing components such as phenols.

In many countries there is an increasing legal requirement for covered water treatment plants to avoid odor and for total nitrogen removal for further protection of surface waters (rivers, lakes, etc.). The latter usually requires an additional purification stage with increased residence time. After the biological stage, the water is clean and can be discharged into the receiving water. The average analyses of the wastewate

COD	$60 - 100$
BOD	$5 - 15$
Oil	$0.5 - 2$
Settleable solids	$0.1 - 0.3$
Phenols	$0.1 - 0.2$

r from a modern refinery (in mg/L)

are:

7.13   HYDROCARBONS IN SOIL AND GROUNDWATER

Since hydrocarbons are water-soluble (even if only to a small degree), their penetration into the soil with possible contamination of the groundwater must be carefully avoided, whenever crude oil or any of its products are handled. Transport from the oil terminal to the refinery is carried out almost exclusively in underground pipelines, which are also the safest means of transport.

The choice of high-grade steels as construction material, good insulation, cathodic corrosion protection, and continuous monitoring for leaks, including visual monitoring from aircraft or ground inspection, ensure a high level of safety. On difficult terrain and in areas of extreme temperature fluctuations, additional measures must be taken (pipeline compensation, elevated piles, intermediate tanks etc.).

The location of the refinery must be carefully selected for possible dangers to drinking water. According to new legislation, all HC-handling units must be erected. That is, to prevent the discharge of spilled product to the underground, to adjacent streets, or canals; the storage tanks are placed in collection spaces which are made impermeable to oil using clay layers, plastic tilts, or concrete lining; in case of a leak they must be capable of receiving the entire tank contents.

If hydrocarbon contamination occurs in the soil, the affected portion of the soil must be removed to prevent subsequent pollution of groundwater. If small amounts have escaped, the contaminated soil is usually combusted in incinerating plants. With larger amounts, and particularly if large areas are polluted with chemical residues and dangerous refuse, the damage must be treated in situ. Depending on the nature of the soil and the corresponding migration of the oil, this can be done by pumping off and purifying the contaminated water, possibly by additional injection of fresh water in adjacent wells. Degradation of the oil by microorganisms is becoming increasingly important. This can also be done in situ or by excavating the soil and treating it externally.

Loading and storage of the products outside the refinery are subject to similar regulations. Due to the large number of external distribution depots for transport fuels and heating oils and service stations, and because of the enormous number of oil-heated households, special care must be taken against overfilling and escape of products due to leaks. Corrosion-resistant and non-ageing steels, plastic-lined steel tanks, and novel, glass-fibre reinforced, plastic tanks are widely used. These tanks must also be constructed such that the whole volume can be collected if leakage occurs (double-walled tanks).

In some countries with a population density and dense housing and industrial areas, the regulations that apply to pure hydrocarbons (i.e., products) are also applied to the handling of process water with much lower HC concentrations.

Sulfur and Nitrogen Compounds:

As a natural product, crude oil also contains heteroatomic compounds containing sulfur, nitrogen, and oxygen in addition to hydrocarbons. Whereas the nitrogen and oxygen contents are in the ppm range and play only a secondary role in atmospheric emissions, the sulfur content of the crudes can be as high as several percent. The distribution over the individual refinery fractions varies, but the content increases with increasing molecular size.

Sulphur Compounds: Sulphur and its compounds are catalyst poisons and adversely affect atmospheric emissions. Hydrogen sulfide, mercaptans, and disulfides are odor nuisances and sulfur dioxide is formed during the combustion of crude oil products. Therefore, sulfur and its compounds must be removed or their contents reduced. The light refinery products, liquefied petroleum gas, and gasoline must be almost completely sulfur-free; for diesel fuels and light heating oils, a substantial sulfur reduction of 0.1–0.5% is required by recent legislation.

Serious problems exist with heavy fuel oil, which is used almost exclusively as fuel in large industrial furnaces and power stations and leads to considerable SO2 emissions. Many countries have established maximum sulfur content in fuels of 1–2wt%. This value can be reached without additional treatment only with a few, low-sulfur crude oils, whose supplies are limited.

Sulfur is generally removed from distillates by hydrodesulfurization, whereby the chemically bound sulfur is converted to hydrogen sulfide. The H2S is then removed in a gas scrubber, converted to elemental sulfur in the downstream Claus process, and supplied to the chemical industry as a raw material.

The intense smell of the H2S-containing gases to be processed and the high toxicity of H2S, even at high dilution, means that precautions must be taken when handling. All sulfur processing plants must be completely gastight: in the hazard zones, instruments, and alarm devices are installed which automatically shut down the plants in case of danger.

CHAPTER 8

SAFETY AT WORK

8.1   THE OCCUPATIONAL SAFETY AND HEALTH ACT

Occupational health hazards are those factors arising in or from the occupational environment that adversely affect health. Thus, the Occupational Safety and Health Administration (OSHA) came into being in 1970 and is responsible for administering the Occupational Safety and Health Act. The goal of the Act is to ensure that employees do not suffer material impairment of health or functional capacity due to lifetime occupational exposure to chemicals. The statute imposes a duty on employers to provide employees with a safe workplace environment, free of known hazards that may cause death or serious bodily injury.

The Act is also responsible for how chemicals are contained. Workplaces are inspected to ensure compliance and enforcement of applicable standards under the Act. In keeping with the nature of the Act, there is also a series of standard tests relating to occupational health and safety as well as the general recognition of health hazards in the workplace. The Act is also how guidelines have evolved for the management and disposition of chemicals used in chemical laboratories.

8.2   FIRE SAFETY

Fire is one of the most destructive, disruptive, and costly causes of damage to any building or oil and gas platform. Fire can be responsible for the loss of jobs, loss of businesses, and loss of life as well as serious damage to the environment. Many companies go out of business following a serious fire. Fires do not just happen they are caused. While many fires start as acts of carelessness, ignorance, or failure to take account of obvious hazards, some are started deliberately. Most fires are preventable by simple precautions and those that do start can usually be held in check or quickly controlled by fire safety measures. These measures can be incorporated into buildings either during construction or renovation works but in addition, well-trained staff can play an equally important role in preventing and tackling fires.

Fire safety is everyone's responsibility. Therefore, all parties must take all reasonable steps to co-ordinate and co-operate with each other to ensure the preventative and protective measures are undertaken for their individual and collective safety. The introduction of Fire Precautions (Workplace) Regulations has introduced the need for a formal fire risk assessment for most places of work. This regulation requires the owners and occupiers of buildings to take a more proactive role in identifying the most likely risks of a fire starting and of the likely risks to staff. It then requires them to take measures to reduce these risks.

In an organization, all employees, interns, contractors, and visitors are required to take reasonable care for the safety of their selves and that of others to achieve the highest standards of fire safety. This includes minimizing the amounts of combustible materials present in buildings, ensuring the safe use, storage, and disposal of flammable substances, and complying with the smoking policy.

8.2.1 What is Fire?

Fire, or combustion, is a chemical reaction in which a substance reacts with the oxygen in the air and emits heat and light.

8.2.2 The Fire Triangle

For combustion to occur, three things need to be present – oxygen (usually from the air); a fuel which can either be a liquid, a gas, or a solid; and heat from some external source. These three elements are usually represented graphically as the fire triangle (Figure 9).

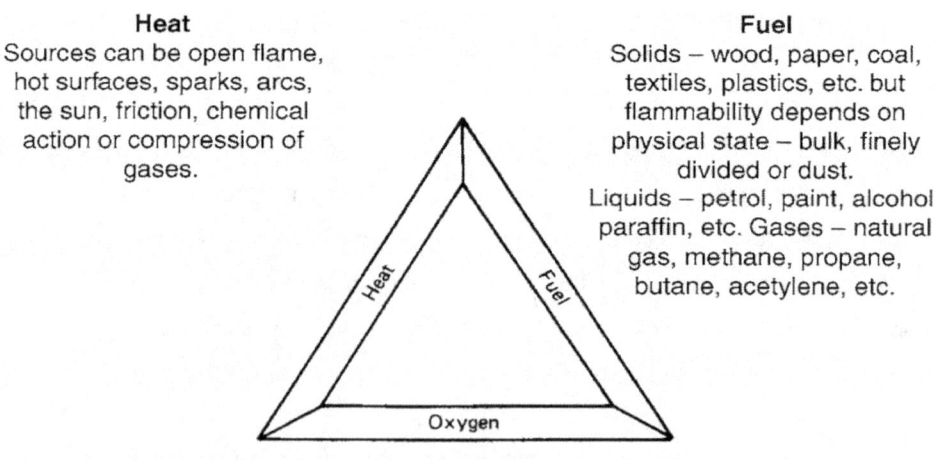

Heat
Sources can be open flame, hot surfaces, sparks, arcs, the sun, friction, chemical action or compression of gases.

Fuel
Solids – wood, paper, coal, textiles, plastics, etc. but flammability depends on physical state – bulk, finely divided or dust.
Liquids – petrol, paint, alcohol paraffin, etc. Gases – natural gas, methane, propane, butane, acetylene, etc.

Oxygen
Approximatley 16% required for combustion. Normal air contains 21%. Some substances contain sufficient oxygen to support combustion.

Figure 8: The fire triangle

8.2.2.1   Fuel

The flammability of solids depends on their physical state. Usually, the more finely divided they are, the more rapidly they will burn, e.g. sawdust will burn quicker than a tree trunk, and coal dust more quickly than large lumps of coal. Solids and liquids do not burn but when heated, give off a gaseous substance or vapor that burns. For example, coal when heated gives off methane, benzene, and other hydrocarbon gases and it is these gases which burn rather than the coal itself. Flammable liquids always have a gaseous layer above their surface and it is the layer that burns and not the actual liquid.

8.2.2.2 Oxygen

The main source of oxygen is from the air, which contains 21% oxygen. The level of oxygen must be reduced to below 16% for combustion not to occur or be suppressed. Some substances, such as chlorates and organic peroxides, contain sufficient oxygen in their chemical makeup to sustain combustion without any additional oxygen being available.

8.2.2.3   Heat

The most usual source of heat that provides sufficient energy for combustion to commence is external sources such as a match, heat from friction or electrical short-circuits. Other forms of ignition include a spark from static electricity or making and breaking of an electrical contact.

8.2.3   Spontaneous Combustion

Spontaneous combustion occurs in materials where some internal chemical or biological action causes a temperature rise sufficient to ignite the material. Other materials can react with oxygen at normal temperatures to generate heat. If the heat is not able to dissipate quickly enough then the temperature of the material will continue to increase until eventually it reaches its ignition point when it will ignite. Most organic materials such as coal, hay, straw, wood, or paper, are prone to spontaneous combustion when stored in bulk, particularly if damp when initially stored. Such as chlorates, when stored in bulk, are prone to explosive ignition with disastrous results.

The risk of spontaneous combustion of such materials can be reduced by ensuring they are dry when stored, by providing sufficient circulation of air through the storage area, and by careful monitoring of the temperature build-up during storage.

8.2.4   Spontaneous Ignition Temperature

A material will ignite spontaneously at this temperature. Some materials have so low an ignition temperature that if exposed to normal room temperatures the substance will ignite spontaneously without the need for an external source of heat or flame. This is referred to as auto-ignition and the temperature at which it occurs the auto-ignition temperature. Solids that exhibit this characteristic include white phosphorus, which must be kept immersed in a liquid to exclude it from contact with the air.

8.2.5   Smouldering

Smoldering is a very slow combustion in the air and can exist for long periods without any noticeable flame. It can occur in porous materials such as paper, sawdust, and latex rubber and produces large volumes of flammable smoke, which accumulates until it reaches its lower flammability limit when it ignites. Smouldering can be transformed into a flaming fire if the supply of oxygen increases. An example of smoldering combustion occurs in upholstered furniture, ignited by a cigarette, which can lie dormant for a considerable time before bursting into flame.

8.2.6   Fire Spread

Fire can spread through a building once it starts, and there is sufficient fuel and oxygen to sustain it in three ways. They are by conduction, convection, or radiation.

8.2.6.1   Conduction

Conduction, although most evident in solids, can occur also in liquids or gases. Conduction is the carrying of heat through or along a material. Heat energy is passed from molecule to molecule and flows away from the source of heat towards areas of lower temperature. The ability to conduct heat varies between materials. In general, good conductors of heat are also good conductors of electricity, i.e. metals. In a fire, a metal girder passing through a fire compartment wall may conduct enough heat to ignite materials in the neighboring compartment. A steel door with no insulation will conduct heat much better than a wooden door although initially, it may resist a fire better.

8.2.6.2   Convection

Convection is the carrying of heat by the internal movement of molecules within a material and only occurs in liquids and gases. It occurs when water in a saucepan is heated. As the water at the bottom of the saucepan heats up it becomes less dense and rises with the colder denser water taking its place at the bottom of the pan. A similar effect occurs in a fire with the hot air or smoke rising and cooler air being drawn in at the base of the fire. The rising heated smoke forms a plume until it reaches a horizontal surface such as a ceiling where it spreads out. As the smoke gets further from the fire, it cools and drops towards the floor forming a 'mushroom effect'. As this process continues, the temperature of the plume will gradually rise until it reaches a temperature at which it will ignite any combustible materials with which it comes into contact.

8.2.6.3   Radiation

Radiation is the emission of rays that transfer heat energy through the atmosphere. It will heat solids and liquids but not gases. It does not involve any physical contact between the source and the target material. For example, heat energy from the sun travels through space and the earth's atmosphere to warm the surface of the earth. Heat is radiated as infrared radiation and its transfer can be likened to the transfer of light. Radiated heat can pass through some materials such as glass, and ignite combustibles on the other side. Other radiations occur as electromagnetic radiations where the transmitter emits electromagnetic waves that act directly on the molecules of a solid or liquid increasing their energy and hence raising the temperature of the material. Microwave ovens and radio frequency welding equipment emit radiations of this sort.

8.3   FIRE HAZARDS AND THEIR CONTROL

There are several causes of fires but the most common ones and some basic precautions to prevent them are highlighted here. The most common causes of fire are arson, combustible dust, electricity, smoking, hot work, heating systems, housekeeping, and lighting.

8.3.1 Arson

Arson is the single largest cause of fires in the workplace. Those people most likely to commit arson are:

i.   People with a grudge against the company or individuals within it;

ii. Intruders to destroy evidence of another crime;

iii. Staff to cover a fraud from which they have gained financially;

iv. Members of action groups campaigning against particular products or practices;

v.   Vandals, usually children or youths who are opportunists rather than deliberate arsonists;

vi. Arsonists who enjoy the spectacle of a fire.

Several measures can be taken to prevent arson and these include:

o   Security arrangements to control visitors entering a premises;

o   Securing all windows and doors at the end of the working day;

o   Good perimeter fencing and external lighting with CCTV coverage;

o   Good housekeeping to prevent the build-up of combustible materials and rubbish both inside and outside the building. Skips and other rubbish containers should be located well away from buildings;

o   Careful selection of new employees and pre-employment checks including the following up of references from previous employers.

In addition, should arson occur, the extent of damage can be minimized if an early warning of a fire can be given. This can be achieved by a sprinkler system, which will restrict the spread of the fire, coupled with an automatic fire detection installation linked to a permanently manned control center that can call out the fire brigade.

8.3.2 Combustible dusts

Combustible dust, such as sawdust, flour, cornstarch, etc., in bulk, will only smolder if ignited. However, if the dust occurs as a cloud in the atmosphere, i.e. presents a large oxidizing surface area, it can ignite with explosive force and cause extensive and devastating damage to a building. An explosive concentration of dust would be a health hazard and as such would not be acceptable in a working atmosphere. The normal sequence of events is that a minor explosion occurs in a part of the plant and dislodges dust that has settled on joists, roof trusses, and other parts of the building. This disturbed dust forms the explosive cloud that causes a secondary devastating explosion. Such dust explosions can be prevented by:

- ❖ Enclosing the process plant to prevent the escape of dust;
- ❖ Installing an effective dust extraction system;
- ❖ Maintaining high standards of housekeeping, particularly on surfaces at high levels, i.e. roof trusses and girders;
- ❖ Excluding sources of ignition, such as friction, static electricity, naked flames, and the effective containment of electrical switchgear.

8.3.3   Electricity

Approximately 25% of fires in industrial premises are caused by electricity. A major cause is the overloading of the conductor resulting in its overheating and causing a breakdown of its insulating sheath. This can lead to a short circuit which, in turn, can ignite flammable materials or vapors. Electrical fire from supply cables and wiring can be prevented by ensuring that:

I.   the supply cables, connections, and connectors are in good condition;

II.   the supply cable can meet the electrical demands;

III.   the rating of fuses or circuit breakers that protect individual circuits is appropriate for the current-carrying capacity of the cable;

IV.   all electrical appliances and equipment are inspected and tested at regular intervals;

V.   the condition of supply cables and wiring, especially of portable leads, is inspected and tested at regular intervals;

VI.   all appliances, except double insulated, are effectively earthed;

VII.   appliances and equipment are protected by an RCD or similar earth leakage protection device;

VIII.   The provision of an adequate number of socket outlets or multi-point extensions leads to eliminating the need for socket adapters.

Other electrical causes of fires include:

Arcing is when contactors make or break a circuit. This risk can be reduced by ensuring the contactor loading is within the manufacturer's limits;

Electrostatic discharge. Electrostatic charges can acquire sufficient energy to ignite materials. Static build-up can be prevented or reduced by earthing (grounding) or the use of static eliminators.

8.3.4 Smoking

The number of fires caused by cigarette ends has decreased in recent years, largely through much tighter controls on smoking in the workplace, either a complete ban or through the provision of allocated smoking areas that have been provided with fireproof receptacles for cigarette ends. However, there is a need for constant monitoring to ensure that smoking does not occur in private places such as toilets.

8.3.5 Hot work

Hot work, involving gas cutting and burning, and gas and electric arc welding and burning, is a high fire risk operation and should only take place when suitable fire prevention measures are in place. These should include the provision of local portable extinguishers, suitable fire/spark resistant screening, and, in areas of high fire risk, the issue of a permit-to-work. A trained firefighter should remain in attendance until at least half an hour after the work is completed, and the area should be checked for at least a further hour.

8.3.6 Heating systems

Hot pipes, ductwork carrying high temperatures, and high-pressure substances can pose a high fire risk. They should be lagged particularly where they pass close to combustible materials. Combustible materials should not be stacked or stored against hot pipes but a suitable space to allow circulation of cooling air should be left. Where portable heaters are used in the workplace, they should be tested to ensure that they are in a safe working condition. Portable heaters should be securely mounted or fixed in position since loose heaters can be knocked over with the risk of igniting adjacent materials. They should not be placed near combustible materials when switched on. Heating appliances, whether portable or fixed, that use gas as the heating medium should be isolated at the main supply valve and not rely on the valve on the appliance itself – the flexible supply pipes can become porous and leak gas causing a potentially explosive atmosphere. The siting of portable heaters is important and they should not have product stored in front of them nor should clothing be put on top of them to dry.

8.3.7 Housekeeping

The fuel required for a fire can be provided by rubbish left in the workplace. High standards of housekeeping play an important role in fire prevention. Any rubbish should be cleared up and disposed of in a proper receptacle. If the rubbish is contaminated by a flammable liquid or substance, it should be stored in a fireproof container.

8.3.8 Lighting

Modern fluorescent luminaires do not generate a great deal of heat. However, tungsten lamps, by their nature rely on very high temperatures in the element to create light. The heat from the element can be transmitted to the bulb casing, which can reach temperatures high enough to ignite combustible materials placed close to them. Where tungsten lighting is used, combustible materials should be stowed well clear of them.

8.4 FIRE ALARMS AND DETECTORS

The priority response in the incidence of fire must be the saving of life. To this end, the detection of the outbreak of a fire and the sounding of an alarm play a crucial role. The means of detection and the type of alarm employed will depend on the type of organization, the operations it carries out, and the number of employees at risk. A characteristic of an alarm is that it must be capable of being heard in every area to which an employee may need to have access, including inner offices, stores, toilets, etc.

8.4.1 Manually Operated Fire Alarms

The simplest form of fire alarms is manually operated devices such as gongs or bells. These are suitable only for the smallest premises where one of the devices can be heard throughout the whole building. For larger premises, an electrical fire alarm system will be required incorporating break glass call points and sounders (audible warning) such as bells, sirens, or hooters. In a medium-sized building, the alarm system could employ a single circuit but for more complex buildings and plants a multi-zone alarm system may be required that divides the building or site into discrete fire warning areas. This can give a local area evacuation warning and also give a general alert warning to other areas.

Audible messages can also be part of the alarm system particularly where members of the public are present. These will alert the occupants to the fact of a fire and can be used to direct them to the nearest fire exits by the safest routes. Tests have shown that the public reacts more quickly to an audible message than they do to a bell or other sound. The standards required for voice alarms are contained in BS 5839: Part 82 and EN 608493.

8.4.2   Automatic fire alarms

Automatic fire alarms have the advantage of being able to raise the alarm in the event of a fire in an unoccupied or unmanned area. They operate by detecting particular changes in the environment. Different types of detectors have been developed to react to the different stages in the development of a fire. The different characteristics that may be used in detecting a fire include:

- o   variations in the strength of transmitted beams or rays between transmitter and receiver, whether visible light, infra-red or radioactive, caused by the rising hot products of combustion;
- o   the visual interference (obscuration) of a light beam caused by smoke from burning materials;
- o   light from flames impinging on a photo-electric cell, but this may be distorted by the level of local illumination;
- o   changes in temperature levels either to above a pre-set level or by a rate of temperature increase above an ecological norm;
- o   Changes in the reception of existing remote-controlled visual security monitoring systems.

The stage at which fire is detected depends on the type of detector used.

The three most common types are:

- ▪   Smoke detectors;
- ▪   Heat detectors;
- ▪   Flame detectors

8.4.3   Smoke detectors

Smoke is a complex mixture of gases, liquids, and solid particles depending on the material that is burning and the conditions of combustion. Each of the constituents of smoke displays particular optical and physical properties which are exploited in the two main types of detectors – optical and ionization. Aspirating smoke detectors use either optical or ionization detectors. The general requirements for fire alarm smoke detectors are contained in BS EN 544 with detailed requirements for smoke detectors.

Recent research has resulted in the development of carbon monoxide detectors. They are most effective in detecting slow smoldering fires where high levels of carbon monoxide are produced before there is any smoke. They are not intended as replacements for smoke detectors but as an addition to the range.

8.4.4 Ionisation detectors

Ionization detectors contain a radioactive source, which ionizes the air within a containing chamber resulting in a small current flow between two electrodes. Particles of smoke entering the chamber interfere with the ion transport and lead to ion–electron recombination, thus reducing the current flow. This reduction in current is sensed and triggers the alarm. Ionization detectors respond more quickly to smoke containing small particles but have a less rapid response to smoldering fires involving polymers (plastics etc.).

8.4.5 Optical detectors

Optical detectors contain a light-emitting diode and a receiver. Either smoke particles entering the detector obscure or scatter the light beam causing the output signal from the receiver to vary. This in turn triggers the alarm. Optical detectors respond more effectively to dense, heavy particulate smoke such as that generated by oil and plastic fires. Optical detectors are of two main types:

Light scatter type where smoke entering the detector reflects the light from a light source onto a photo-electric cell. The small electrical charge produced by this is amplified and actuates the alarm relay

Obscuration-type detectors are usually installed to span large areas. A light source and lens are positioned at one end of the area and a receiving photoelectric cell is located at the other end. Rising smoke from a fire passes through the beam deflecting the beam or obscuring the light. This results in a reduction in the intensity of light falling onto the photoelectric cell and causes the alarm signal to be triggered

8.4.6 Heat detectors

Heat detectors can be of either fixed temperature, rate of temperature rise, or linear type. Fixed temperature detectors are activated when the temperature in the area reaches a predetermined level and operate similarly to a thermostat. The rate of temperature rise detectors are activated when there is an abnormally rapid increase in the temperature at the detector above that experienced with heating systems or sunlight.

They can incorporate an upper-temperature setting to provide a warning where the rate of temperature rise is below the detectable level. Linear heat detectors comprise a detection cable that runs through the area to be protected. The alarm is raised when the electrical characteristics of the cable change because of a temperature rise. Heat detectors are most effective in areas where there may be smoke or steam under normal conditions i.e. boiler rooms, kitchens, or areas where, in the event of a fire, it can be expected to be a flaming fire with little or no smoke.

8.4.7   Radiation detectors

As well as producing hot gases, fire releases radiant energy in the form of:
- visible light
- infra-red radiation
- Ultra-violet radiation

These forms of energy radiate in waves from their point of origin and the detectors are designed to respond to measurements of this radiation. The use of the visible light band detector has several disadvantages associated with the fact that it is not able to differentiate between the various legitimate sources of visible light and those created by a fire. In practice, radiation detectors are designed to respond to either infrared or ultraviolet radiation. Both of these types of detectors 'look' at the flame and memorize it before having a second 'look' after a short delay to confirm that the flame is still there. If it is, then the alarm signal is triggered.

Infrared detectors can scan large areas and identify the wavelength and pattern of radiation given out by a flame. They can also be used as spark detectors to protect dust extraction systems. Ultra-violet detectors are used in specialized areas such as aircraft engine compartments. Radiation detectors generally complement heat and smoke detectors, especially in tall, unobstructed compartments, and are effective in special applications such as flammable liquid storage areas.

8.4.8   Carbon monoxide detectors

Normal heat and smoke detectors rely on the presence of convection currents caused by the fire to carry either the products of combustion or heat past the detector, which is usually mounted at a high level. If there is little heat from the fire then there can be a significant delay before the detector receives enough information from the fire to actuate the alarm. Carbon monoxide detectors overcome this problem since the carbon monoxide produced by the fire dissipates into the atmosphere of the protected area and can reach the detector without the need for convection currents. This type of detector must not be confused with the carbon monoxide gas sampler used to protect people from the presence of this gas. The two instruments perform quite separate functions with widely different trip settings.

8.4.9 Radio fire alarm systems

Radio alarm systems are especially useful when alarm systems need to be installed into existing buildings that have preservation orders or are listed buildings where any visible disfiguration, such as that caused by wiring, is not permitted. These systems are also suitable for protecting temporary buildings as the system can be installed quickly and is easily removed for use elsewhere when the building is no longer required. The system comprises independent call points, detectors, sounders, and control panels each of which has its power source. They are linked, not by hardwiring but by radio transmissions. Care must be taken to ensure there is no local electromagnetic contamination that may interfere with the system's integrity and that the system's radio signal does not interfere with local, especially emergency service, radio systems.

8.4.10 Control and indicating equipment

The 'heart and brains' of any fire alarm installation, the control and indicating equipment provides the power as well as monitoring the system and indicating the location of any detected fire. The control equipment should be in an easily accessible location so that it can be seen quickly and easily in the event of an alarm and give immediate information on the location of the fire. More advanced systems (known as programmable or intelligent systems) can identify exactly which device has acted and also self-monitor to give early warning of a fault or failure of a component of the system.

8.5   CLASSIFICATION OF FIRE

There are five different categories of fire. These categories and the means for extinguishing the fire are listed in Table 4. The identification of the type of fire is important to ensure the selection of the correct type of fire extinguisher.

Table 4: Classes of fires and suitable type of extinguisher (Ridley and Channing, 2003)

	Water	Carbon dioxide	Dry powder	Foam	Wet chemical	Fire Blanket
Class A Paper, wood	✓	X	?	?	?	?
Class B Flammable liquids	X	?	✓	✓	?	✓
Class C Flammable gases	X	X	✓	X	X	X
Class D Metals	X	X	Special powders only	X	X	X
Class F Deep fat fryers	X	X	X	X	✓	X
Electrical	X	✓	?	X	X	X

✓ Suitable extinguishant
? Can be used but not ideal
X Unsuitable and should not be used

8.5.1   Class A fires

Class A fires include those involving solid materials normally of an organic nature such as wood, paper, natural fibres, etc., in which combustion occurs with the formation of glowing embers. Water is the most effective extinguishing agent and acts by reducing the temperature available to ignite further material.

8.5.2   Class B fires

Class B includes fires involving liquids such as petrol, oil, paints, and liquefiable solids such as fats, waxes, greases, etc. The most effective extinguishing agents are foam and dry powder, which blanket the burning material and exclude oxygen. Cooking oils and fats are excluded from this class and are now designated as class F.

8.5.3   Class C fires

Class C includes fires involving gases such as butane and propane. Extinguishing these fires is by shutting a valve in the supply line, i.e. removing the fuel. If this is not immediately possible, special techniques that require expert knowledge need to be used and should be left to the specialists. Until this can be effected, gas fires should be left burning and the main preventative action should be protecting buildings and property in the surrounding area. If the flames of a gas fire are extinguished before the supply can be isolated, there is a danger of the build-up of an explosive gas mixture.

8.5.4   Class D fires

Class D fires involve metals such as aluminium, magnesium and sodium. Most metal fires are difficult to extinguish as they burn at very high temperatures and react violently with oxygen, either in the air or in the extinguishing medium. In the reaction, the oxygen is removed from the water releasing hydrogen, which then ignites violently. Metal fires will normally only respond to the correct extinguishing medium for the type of fire. Wherever it is safe to do, so, metal fires should be allowed to burn themselves out with the surrounding area being protected from fire spread. The use of water or allowing the fire to burn out will not be possible where metal fires occur within a building. Suitable extinguishing substances, such as powdered graphite, powdered talc, soda ash, limestone and dry sand, should be used. These must be applied gently to form a coating and hence smother the fire. Premises where a risk assessment has identified metal fires as one of the hazards should have suitable extinguishing medium available and staff specially trained in the techniques of dealing with this type of fire.

8.5.5   Class F fires

Class F includes fires involving commercial deep fat and oil fryers. These fires have been given a separate classification because the high temperatures involved make the inclusion of these substances in class B inappropriate. Most of the extinguishers suitable for class B fires would not extinguish fire involving cooking oil. A special wet chemical extinguisher has been developed which cools and smothers the fire by emulsifying the cooking oil and sealing its surface with a non-combustible crust to prevent re-ignition.

8.5.6   Electrical fires

There is no separate classification for electrical fires since electricity, while being a cause of fires, is not a flammable material itself. It is important in any fire involving live electrical equipment that the electrical supply is isolated before attempting to tackle the blaze. If that is not possible, carbon dioxide or dry powder extinguishers only should be used. Once the supply has been isolated, the extinguisher appropriate to the burning material can be used.

www.ingramcontent.com/pod-product-compliance
Lightning Source LLC
Chambersburg PA
CBHW071053290526
45795CB00004B/1476